ENGINEERING EDUCATION

Aims & Goals for the Eighties

An Engineering Foundation Conference held
July 26-31, 1981
Franklin Pierce College
Rindge, New Hampshire

CHAIRMAN
William H. Corcoran
California Institute of Technology

CO-CHAIRMEN
Leland J. Walker
Northern Testing Laboratories

David R. Reyes-Guerra
Accreditation Board for Engineering & Technology

Director of Conferences
Sandford S. Cole

SPONSORED BY
Engineering Foundation
Accreditation Board for Engineering & Technology

CO-SPONSORED BY
National Academy of Engineering
American Society for Engineering Education
Education Affairs Council of AAES

PARTICIPATING BODIES OF ABET

American Congress on Surveying and Mapping
American Institute of Aeronautics and Astronautics, Inc.
American Institute of Chemical Engineers
American Institute of Industrial Engineers, Inc.
American Institute of Mining, Metallurgical and Petroleum Engineers
American Nuclear Society
American Society of Agricultural Engineers
American Society of Civil Engineers
American Society for Engineering Education
American Society of Heating, Refrigerating and Air-Conditioning Engineers, Inc.
The American Society of Mechanical Engineers
The Institute of Electrical and Electronics Engineers, Inc.
National Council of Engineering Examiners
National Institute of Ceramic Engineers
National Society of Professional Engineers
Society of Automotive Engineers
Society of Manufacturing Engineers

MEMBER BODIES
American Academy of Environmental Engineers
American Society for Metals

Any findings, opinions recommendations, or conclusions contained herein
are those of the individuals, and do not necessarily reflect the opinion of,
nor imply endorsement by, the sponsors or co-sponsors.

Copyright 1982 United Engineering Trustees, Inc.
ISBN 0-939204-13-4
Library of Congress Catalogue Card No: 82-71877

Available from: Accreditation Board for Engineering & Technology
                  345 East 47th Street
                  New York, NY 10017

INTRODUCTION

For the last three years, there has been a perceived crisis generating in engineering education. Several groups have been and are currently examining engineering education with an emphasis on critical issues. Factors which have been cited as possible areas of concern are: (a) increased enrollments (an 89% increase in freshmen enrollments over the last five years), (b) lack of qualified and available faculty, (c) a decrease in graduate students who are U.S. citizens; (e) loss of research support, (f) inadequate and deteriorating laboratory facilities and equipment; and (g) higher-then-ever beginning salaries for engineering graduates.

Data on each of the areas of concern have been difficult to obtain (qualitative information is available), but quantitative information is not generated easily. The reports on engineering education show a divergence of opinion as to which areas are critical. Therefore the Accreditation Board for Engineering and Technology thought that an intensive discussion among representatives from universities, industry, government and foundations could lead to more integrated thinking on major current and future issues in engineering education.

The Engineering Foundation supported ABET's proposal for a conference. Co-sponsorship was offered to and accepted by the National Academy of Engineering, the American Society for Engineering Education and the Educational Affairs Council of the American Association of Engineering Societies. The conference titled "Engineering Education - Aims and Goals for the Eighties" was held on July 26-31, 1981, at Franklin Pierce College in Rindge, New Hampshire. There were 128 participants.

In order to provide reasonable order for the program, each day was devoted to a broad topic. On Monday, July 27, the theme was pre-college experience for engineering education. On Tuesday, July 28, it was engineering education and the baccalaureate experience. The general topic for Wednesday, July 29, was graduate education in engineering. Thursday, July 30, was spent on engineering practice and engineering education. On the morning of Friday, July 31, a summary and recommendations were developed based upon the discussions held during the week.

The intent of the meeting was not to provide the last word on engineering education and aims and goals for the eighties, but rather to focus upon key issues. The thought was that with such focus, appropriate conclusions could be developed to provide foundations for further discussions, analysis, and action by the various groups in the United States currently concerned with problems in engineering education. In addition, a goal of the conference was to develop sufficiently concise conclusions and recommendations to be of value to President Reagan and the Executive Branch of the government as they ponder actions relative to the technological future of the United States and the strength of its educational processes, especially in engineering.

Because the goal of the conference was to communicate as described, the lead presentation in this report gives conclusions and recommendations. After that section, an executive summary is presented that gives more detail on the discussions that were held. The third section of the report presents the supporting papers and documents developed during the course of the meeting. Finally there are appendices with lists of attendees and the planning committee for the meeting.

In order to expedite dissemination of this report, several versions have been prepared. A copy of the total report can be obtained from ABET headquarters, 345 East 47th Street, New York, New York 10017.

The conference identified a number of areas which affect engineering education. ABET, as the body responsible for the maintenance of quality in engineering education through its accreditation process, is carefully studying the conclusions and recommendations resulting from the conference. Emphasis will be given to many of the issues through the accreditation process.

This report is comprehensive and ABET offers the conclusions and recommendations in the spirit of assisting all those who have an interest in engineering education. Many of the topics presented should be further expanded and addressed by other organizations. ABET has assembled the report and if there are points that are hazy, controversial, or even incorrect, that result is the responsibility of ABET. If an effort were made to reach exact agreement on all points discussed at the conference, a report most probably would not be forthcoming. The presentation is close to a consensus. With that warning and disclaimer, the report has been prepared. The content should indeed provide stimuli for other groups analyzing engineering education and should be of special value to the government of the United States as it proceeds to deal with critical problems in engineering education.

TABLE OF CONTENTS

INTRODUCTION iii
CONTENTS v
CONCLUSIONS & RECOMMENDATIONS 1
GENERAL CONCLUSIONS & RECOMMENDATIONS 3

EXECUTIVE SUMMARY OF SESSIONS
    The Pre-College Experience for Engineering Education
        David R. Reyes-Guerra 19
    Engineering Education and the Baccalaureate Experience
        Russell R. O'Neill 21
    Graduate Education in Engineering
        Robert H. Page 25
    Engineering Practice and Engineering Education
        Charles E. Schaffner 29
    Research and Engineering Education
        Richard G. Cunningham 35

OPENING REMARKS
    Leland J. Walker, President, ABET 39

CONFERENCE PAPERS
    Opportunities for the Gifted Student with Engineering Interests
        Lee F. Browne 41
    Engineering Education and the Baccalaurate Experience
        O. Allan Gianniny 51
    Effect of Progress in Microelectronics in Education in Electrical Engineering
    Especially and Engineering Education in General
        Louis T. Rader 53
    Teaching of Engineering Courses with Student Faculty Ratios Far in
    Excess of Past Experiences
        John C. Lindenlaub 65
    Development of Clear, Easy Writing and Clear, Plain Speaking
        Ralph Jenkins 69
    Key Aspects of Comparison between Engineering Education in USA and
    England, Japan, Germany and Russia
        J. S. Przemieniecki 77
    Crisis in American Engineering Education
        Jerrier A. Haddad 89

How to Increase the Number of Ph.D. Candidates and
Supply of Faculty Members
  Donald D. Glower 97
Implications of Increasing Percentage of Foreign Nationals Enrolled in
Graduate Programs for Engineering
  W. Robert Marshall 103
Effect of Federal Support for Programs and Students on Generation of
New Faulty Members
  Ross J. Maritn 109
Concept of a Professional School of Engineering
  Charles H. Samson 117
The Need for Continuing Education and Its Quality Control
  Roy H. Mattson 137
Effect of Computer Graphics on Both Industry and Engineering Education
  Edward M. Rosen 139
Education for Engineering Management in the Eighties
  Merritt A. Williamson
  (paper submitted but not presented by author) 157

APPENDIX I 167

APPENDIX II 171

APPENDIX III 179

AUTHOR INDEX 183

SUBJECT INDEX 185

CONCLUSIONS AND RECOMMENDATIONS

What follows is a presentation of conclusions and recommendations developed in the five-day meeting at Franklin Pierce College. A stepping back from the tabulation gives perspective on the two prime critical issues in engineering education. They are repeated here at the beginning of the summary in briefer form to emphasize their importance in our coping successfully with the future of engineering education in the United States. They are:

First Prime Critical Issue

Conclusion: The ratio of students to faculty members in engineering schools in the United States has increased to a dangerous level. Quality of engineering education is decreasing where there has not been at least a concomitant increase in budget for teaching assistants and support services.

Recommendation for action: If quality education is to be maintained we must either

(a) limit enrollments, or

(b) increase faculty size, or

(c) increase support staff, including graduate teaching assistants, or

(d) act with combinations of (a), (b), and (c) above

Second Prime Critical Issue

Conclusion: Insufficient numbers of baccalaureate degree students that are U.S. citizens are entering graduate school in engineering. They are choosing industrial careers instead. The result will be a decrease in the percentage of U.S. citizens who become engineering professors and a decrease in our long-term creativity and productivity in high-technology areas to the extent those areas are affected by proper numbers of Ph.D.'s entering those fields each year.

Recommendation for action: We must

(a) Make graduate study more attractive by increasing current stipends from their current level to annual levels equal to one-half of the entry level salaries for baccalaureate degree graduates entering industry. Immediate measures needed to increase current stipends of $7000 annually to $11,500.

(b) Make the teaching profession as economically rewarding as work in industry. As soon as possible, the annual starting salaries for assistant professors should be increased to a level equal to one-third more than the entry level salaries for baccalaureate degree graduates entering industry.

GENERAL CONCLUSIONS AND RECOMMENDATIONS

In addition to the two prime critical issues, ABET also agrees that there is another major issue - the obsolescence and lack of adequate laboratories, including computer facilities, as well as instruction and support personnel for the laboratories. ABET will emphasize these areas in its accreditation process.

We must provide, through grants from government and industry, the necessary support to upgrade laboratories. Government can further help by instituting tax incentives and other measures that industry can apply towards their grants in support of laboratory equipment. Additionally, the teaching profession must recognize the need for greater professional academic recognition and rewards for laboratory instruction.

The complete summary of conclusions and recommendations for action is presented in this section. There has been no attempt to establish priority. The topics follow the agenda of the meeting.

I. **THE PRECOLLEGE EXPERIENCE IN ENGINEERING EDUCATION**

A. Conclusion: Technological literacy needs to be improved from first through twelfth grades.

Recommendation for Action (hereafter RFA):

Professional engineering societies should note the presence of strong organizations such as JETS, MESA, PRIME and cooperate with them in communication with grammar schools and junior high schools, at least, on current technical problems.
Industries should provide case studies on application of technical knowledge. They should be used by teachers, professional societies, and youth oriented organizations in their cooperation to improve technical literacy in primary and secondary schools.
Government should note the special opportunities for increased interaction among industry, universities, and professional societies in communicating technical knowledge to schools.
The State and Federal Government should provide seed grants to encourage the cooperative efforts of professional societies, universities, industry, and government in communication with the primary

and secondary schools on improving their understanding of technology.

B. Conclusion: More role models are needed in the primary and secondary schools in order to allow students to consider engineering as a vocation.

RFA: If the interaction on technical literacy described in conclusion IA is developed, sufficient role models will appear in the structure of the primary and secondary school systems. Major attention must be given to the issue, however, by professional societies, universities, industry, and the government.

C. Conclusion: Teachers in both primary and secondary schools need increased opportunities to learn about modern technical subjects.

RFA: Many approaches to this problem have been used, and a recommendation is made that emphasis be given to the interaction described in RFA IA and IB and efforts made to build upon the results in the most efficient manner.

D. Conclusion: Salaries for primary and secondary school teachers must be increased in order to decrease the drain of qualified teachers to the industrial world.

RFA: Because the problem is national, the Federal Government must support the state and local governments in examination of the improved salaries for teachers if technological growth in the United States is to continue as it should for the quality of our life and the safety and the future of our country. Teachers of technical courses in primary and secondary schools must be technically prepared and their salaries should reflect that particular effort and should have parity with industrial jobs of related years of experience.

E. Conclusion: Students in primary and secondary schools should have increased opportunities to learn about relationships among technological change, productivity, marketing and cost of doing business.

RFA: Throughout the United States, industry should cooperate with the professional societies and organizations such as MESA in communicating such information to school districts without preempting the roles of the schools.

F. Conclusion: The challenge and difficulty of technical courses in both primary and secondary schools must be increased in order to challenge the gifted students

General Conclusions & Recommendations

as well as to provide strong backgrounds for all students.

RFA: Universities should consider increased entrance requirements in technical areas for students planning to enter science and engineering. Concomitantly where high-school students because of economic disadvantage or other problems have less exposure to strong technical programs, universities must take this lack into account and improve the students' opportunities for better educational experiences.

G. Conclusion: Communication with minority students on engineering must be increased in both primary and secondary schools, with emphasis on the primary schools.

RFA: Professional societies must interact with greater strength with active bodies such as JETS, MESA, PRIME, SECME, in order to communicate with the students in the primary and secondary schools. Universities must increase interaction between their own students and the minority students in the primary and secondary schools. The model of the student in the university is a very important model for the primary and secondary school students to see.

II. ENGINEERING EDUCATION AND THE BACCALAUREATE EXPERIENCE

A. Conclusion: With appropriate strength in humanities and social sciences in the engineering curriculum, that curriculum is a liberating education.

RFA: Improved interaction among colleges in a university, universities, accreditation groups, and professional societies is required for better understanding of the technical person.

B. Conclusion: In the baccalaureate experience of the engineer, there should be increased effort in the core program of mathematics, physics, and chemistry to show more about the application of this basic knowledge to real problems.

RFA: Examine Northwestern University's and Purdue's integrated program for the freshman year and determine the practicality of improved interaction among professors of chemistry, physics, mathematics and engineering schools in teaching the basic science courses.

C. Conclusion: Educational progress in microelectronics is basic to the future economic growth and safety of the

ENGINEERING EDUCATION

United States.

RFA:
(1) Undergraduate programs in engineering in all disciplines should include education on microprocessors, preferably early in the curriculum.

(2) Computer-aided design (CAD) and computer-aided manufacturing (CAM) are a part of the industrial scene now. They must currently be a part of the undergraduate education, and increased interaction between the university curriculum and industry to speed educational applications of CAD and CAM must be established.

(3) Universities need new equipment and methods to work on very large-scale integrated circuits and to work on modern design. A minimum sum of $3 million to each of several schools should be allotted by the Federal Government with an additional 15 percent per year being assigned to stimulate the development of integrated-circuit equipment of various types.

(4) The university must pay significant attention to long-term industrial needs, and industry must pay attention to the short-term academic needs in application and understanding of microelectronics.

(5) Increased understanding of the role of microelectronics in improved productivity must be developed in the baccalaureate program by improved interaction between industry and the university.

(6) Where possible, state legislatures should appropriate money for state universities to improve the equipment available for education and research in microelectronics.

(7) Effort should be made among the universities to increase interaction of the professors working in microelectronics.

(8) The Accreditation Board for Engineering and Technology should emphasize the value of microelectronics in the undergraduate programs and press for increased financial support, especially from state governments.

(9) Professional licensing boards should increase the amount of microelectronics knowledge

General Conclusions & Recommendations

required on professional engineering examinations.

(10) An increase in microprocessor designers above the current number of 2,250 should be developed, especially by way of continuing-education courses.

(11) An increase in consortia, such as that involving cooperation of the University of Virginia and other universities in that geographical area interested in development of microprocessor curricula and research, should be encouraged by state governments, the Federal Government, professional societies, and the Accreditation Board for Engineering and Technology.

(12) There should be a combination of microcomputers and video discs to improve education at the undergraduate level.

(13) Improved continuing education should be made available to older professors to get them up to speed in the technology of microelectronics One computer company is already providing such support.

(14) Individual states should evaluate resources for computer-aided design, and education of engineers, engineering technologists, and engineering technicians should be integrated in the approach to understanding computer-aided design and its application.

D. Conclusion: Perhaps the key critical issue facing engineering education in the United States today is the increasing ratio of students to faculty in engineering curricula throughout the country. That increased ratio has not concomitantly seen an increase in support services providing added graduate teaching assistants and other elements needed for continual improvement of the educational process.

RFA: (1) The Accreditation Board for Engineering and Technology must pay closer attention to the quality of student work presented by a university in the accreditation of an engineering program. Emphasis should be placed upon communication with students in evaluating that work in the presence of the increased student-faculty ratios.

(2) In the absence of appropriate funds to maintain the quality of engineering education, engineering enrollments must be restricted. More than 30% of the engineering schools in the United States today are restricting enrollment. That restriction has to be balanced against the needs of the United States in the future for engineering teachers and for practitioners in the development of high technology so important to our productivity and national well being.

(3) In order to maintain quality in the face of decreasing dollars available per student, special note should be made of the opportunity for use of programmed instruction in basic engineering education and increased use of experienced engineers in the teaching of synthesis and design. These changes, nevertheless, have to be considered in the light of significant constraints on budgets.

E. <u>Conclusion</u>: Even though we are seeing an increased number of minorities entering engineering in the universities in the United States we are not seeing sufficient retention of the minorities in the engineering programs.

<u>RFA</u>:

(1) Recruitment of minorities into engineering must be considered in concert with programs for improving retention.

(2) In all aspects of dealing with recruitment and retention, the positive aspects of engineering careers must be emphasized, and the difficulties and the stress in engineering education must be clearly understood at all times. There must be capitalization on the understanding of job opportunities.

(3) Programs for retention of minorities must have institutional commitment, especially from the President of the institution.

(4) Faculty members must understand the problems with retention of minorities, and the maximizing of retention must not be left to random chance.

(5) Be sure that minority students have program administrators with whom they can speak at all time about their problems.

(6) To improve retention a better understanding of the university system must be communicated

to entering minority students.

(7) Strong communication between minority students and faculty members must be sustained.

(8) Engineering societies must help the universities in aiding the minority groups to understand more about the opportunities in the engineering profession. The difficulties in the engineering programs must be matched against the rewards of being an engineer.

F. Conclusion. There must be a renewal of emphasis on the quality of undergraduate engineering laboratories with respect to both instruction and equipment.

RFA: (1) Professors should be given the same instructional credit for laboratories as for classroom instruction because the communication with the student in the laboratory is every bit as equal as the communication in the classroom.

(2) Care should be exercised not to depend totally upon donated equipment because one of the current problems in engineering laboratories in universities in the United States is the presence of obsolescent equipment and insufficient funds to keep equipment up to date.

(3) Engineering professors in each discipline should be totally committed to improving the laboratory operations, with increased emphasis on instrumentation and control of systems.

G. Conclusion: Few universities seriously assume responsibility for development of clear writing and clear speaking on the part of engineering students.

RFA: (1) Fund studies on improved teaching on writing and speaking. These funds should be developed from state and federal government, foundations, and industry.

(2) Consider technical communication as a special aspect of the broad area of communication through writing and speaking.

(3) Pay increasing attention to the needs of students from different cultures in the development of clear writing and speaking.

(4) Help students practice communication skills under real circumstances.

(5) Increase interaction between industry communicators and teachers of writing and speaking to improve communication instruction and practice in the university.

H. Conclusion: To maintain the lead in engineering education, the United States must examine more carefully the nature of engineering education in the United Kingdom, Germany, Japan, and the USSR.

RFA: (1) Engineering curricula in Europe and Japan should be studied to guide further development of engineering education in the United States. These studies can be funded by way of foundations and the federal government.

(2) Consideration should be given to the value of higher-level doctorate degrees in engineering such as given in the United Kingdom, Germany, and the USSR. The purpose would be to determine whether the doctorate beyond the Ph.D. degree would increase focus on basic research and help in the development of increased productivity.

III. GRADUATE EDUCATION IN ENGINEERING

A. Conclusion: A career in the teaching of engineering has decreased in attractiveness and thus raises the question of how we will satisfactorily develop our engineering professors and high technology academic studies in the near future.

RFA: (1) A jump in engineering faculty salaries is required so that at a minimum the starting salary for an assistant professor in engineering is one-third more on an annual basis than the starting salary of the holder of a baccalaureate degree entering the profession of engineering in industry.

(2) Improve the support services and equipment in the engineering schools in order to improve the attractiveness of the surroundings relative to industrial laboratories and operations.

(3) Increase graduate stipends from the current levels of approximately $7,000 per academic year to $11,500 per academic year in 1981 dollars.

(4) Help handle the increased costs in meeting the recommendations by differentially increasing the cost of tuition in the engineering school.

General Conclusions & Recommendations

B. Conclusion: Because of the various adverse pressures on selection by U.S. nationals of graduate school and then teaching as a career, there is an increasing percent of foreign nationals entering graduate schools of engineering.

RFA: (1) If more foreign nationals will be entering teaching in the United States, increased care should be exercised in the graduate program to make certain that the quality of oral and written communication increases or at least remains equal to the current communication in engineering schools.

(2) Foreign students who enter graduate school should be treated as assets not only as potential teachers in engineering in the United States but as teachers of engineering in their own countries.

(3) Because of differences in preparation, care should be exercised that the design content of engineering curricula is not minimized in favor of engineering science by students who have had baccalaureate preparation in other countries.

C. Conclusion: An increase of 25 percent in current research funding would help release time for professors in undergraduate education to help in graduate education.

RFA: (1) Both the federal government and industry should combine efforts to increase research funding in engineering programs in order to provide more released time to improve the graduate experience in engineering education.

(2) An effort should be made to encourage 20-25 percent of the current 60,000 annual recipients of B.S. degrees to opt for Ph.D. programs and approximately 50 percent of the 60,000 to opt for M.S. programs in engineering in order to assure productivity increases in the United States over the next decade and to assure sufficient faculty to prepare people for high technology work.

(3) Provide at least $2,000 per year per engineering student for equipment modification and purchase to improve graduate education in engineering.

D. Conclusion: New types of equipment are needed for research on computer-aided design and computer-aided manufacturing in graduate studies.

RFA: (1) Make ARPA network available for activities such as study and research on integrated-circuit design.

(2) Have professional societies encourage the federal government to continue its strong support of basic research at least by way of the National Science Foundation and the Department of Defense.

(3) Make a major increase in the FY 82 Federal budget for equipment for engineering research. This current sum could easily be doubled in order to improve at least the advanced education of engineers in CAD and CAM to prepare for productivity increases and to improve the education of engineering teachers for the future.

(4) Increase university reliance on research programs supported by the private sector such as that of Monsanto and its agreement with the Harvard Medical School and Exxon and its agreement with MIT. Some $200 million in such agreements have been signed over the past year and should be increased in quantity for the health of the country in development of people and ideas at the advanced level.

(5) Be sure that there is no confusion about the need for improved baccalaureate education of engineers and the need for emphasis on graduate education relative to development of high technology and engineering teachers.

E. Conclusion: Increased consideration must be given to the master's degree as the entry level for engineering practice. Integration of the roles of the engineering technologists from the baccalaureate program, the engineering technicians from the associate-degree program, and engineers from baccalaureate and M.S. programs must be carefully examined.

RFA: (1) Increase discussions among universities, industry, and government on the relative values of engineering technicians, engineering technologists, and engineers in engineering practice in industry and government.

(2) Define more exactly the difference between the engineering technologist and the engineer in order to speed the discussions suggested

General Conclusions & Recommendations

under item (1).

(3) Consider accreditation of engineering programs only at the M.S. level.

F. Conclusion: Perhaps the time is not propitious to change, but current examination of the concept of a professional school of engineering is desirable.

RFA:
(1) Continue to examine the nature of the professional engineer in light of increasing efficiency of the engineer in practice when using modern tools of computation.

(2) Increase the interaction among engineering, business, medical and law schools to establish better perspective on the concept of a professional school of engineering.

(3) Examine the professional school of engineering in light of the potential for increased use of engineering technicians and engineering technologists as support groups for the professional engineer, with the idea that there might be fewer engineers per billion dollars of gross national product when backed up by engineering technologists and engineering technicians as part of the support team for the total engineering function.

IV. ENGINEERING PRACTICE AND ENGINEERING EDUCATION

A. Conclusion: Because of rapid increases in new technical knowledge and concepts, continuing education is becoming increasingly important in the life of the professional engineer. Quality control on programs of continuing education must be conducted with skill by the professional societies, the universities, industry and government.

RFA:
(1) Increase the sensitivity of the boards of engineering examiners, the Accreditation Board for Engineering Technology, and professional societies in understanding the role of continuing education and in understanding the need for quality control to protect the consumer, namely, the practicing engineer.

(2) Great encouragement should be given by the professional societies to avoid obsolescence by use of continuing education courses.

(3) Critically examine the requirement by one state for continuing education as a part of

relicensing to determine whether that should be a uniform practice in the United States.

(4) Determine whether ABET acreditation would be a useful part of strength of continuing education programs.

(5) Avoid continuing education programs that are concerned mainly with the development of income for the giver and emphasize that the programs should exist for development of the receiver.

B. Conclusion: Computer graphics are being used extensively in the industrial and government sectors in engineering practice. Their use must be exploited in both undergraduate and graduate programs in engineering.

RFA:
(1) Increase communication among universities on the quality of computer-graphics systems with emphasis on development of low-cost systems that could be personally available to each student in the course of that student's educational program.

(2) Increase communication among universities, industry, government, and professional groups, such as ACM, NCGA, WCGA, and IFIP, on advances in computer-graphics programs.

(3) Increase the awareness of faculty and students in all engineering disciplines about the availability of publications in computer graphics such as Computers and Graphics, Computer Graphics, Computer-Aided Design, Computer Graphics and Image Processing, the Harvard Newsletter, and the SID Journal.

(4) Note that computer graphics must not be considered a frill in engineering education but must be an integral part of engineering education in all disciplines.

(5) To the acronyms of CAD and CAM must be added the acronyms of CAE, computer-aided engineering. Students should understand that computer-aided engineering is important in developing processes, process accounting and process scheduling.

(6) Communicate with universities on the point that cost is not now a real barrier in the development of computer-aided graphics. APPLE computers, for example, can be used with

General Conclusions & Recommendations 15

graphics equipment for as little as $2,500.

(7) Preparation must be made for personal-computer acquisition as the cost of computer systems for computer graphics continue to decrease. The computer-graphics system will probably be as important to the engineering student in the near years as slide rules were in the 1930's.

(8) A new synthesis in engineering education is needed and would encompass the following: information development; communication, data acquisition; development of data base; and establishment of a network allowing realtime computing, technical writing, searching and sorting, developing algorithms for processing of information, application of information to computer-aided design, computer-aided manufacturing and computer-aided engineering.

(9) Because we are now an information society we must incorporate information processing into our basic engineering programs both at the undergraduate and graduate levels. All engineering students should be comfortable when starting their first professional employment with an assigned terminal for information processing. The best way to develop the comfortableness is early in the engineering education to exploit the opportunities for communication using computer graphics.

(10) Industry is far ahead in the application of computers for information processing, and students and professors alike must develop skills possessed by industrial users.

C. <u>Conclusion</u>: An increasing number of science graduates are being employed by industry and government in engineering roles.

<u>RFA</u>

(1) Care must be exercised to assure that these science graduates in engineering roles have had the opportunity to understand the basics of engineering and of economics in engineering systems.

(2) The scientist employed as an engineer in industry should look to continuing education to continue to expand his or her understanding of engineering systems in contrast to dealing with individual items in a system.

D. <u>Conclusion</u>: Increased numbers of computer networks are becoming available, and desk computers are now joinable to

ENGINEERING EDUCATION

those networks. Desk computers, therefore, become powerful tools in industrial design and engineering education.

RFA:
(1) Consideration must be given to the use of the joinable desk-computers and their influence on relative numbers of engineers and engineering technologists to meet manpower needs in industry and government.

(2) Curricula should be restructured to emphasize the early application in engineering education of the power of desk computers in information processing and in attack of systems problems in engineering.

(3) There should be an increase in access of industrial software in the undergraduate and graduate programs in engineering to improve the opportunities for engineering students to develop early facility with desk computers joined to networks.

(4) Availability of desk terminals will be factors in the accreditation of engineering programs by the mid-1980's, and the Accreditation Board for Engineering and Technology should increase its efforts to encourage the use of these individual terminals and to determine how that encouragement can be implemented.

(5) Because of the nature of desk terminals, opportunities must be made available early in engineering student's careers to improve their typing skills.

(6) Inasmuch as it would appear that high-capacity desk computers will be the avenue for computation and word-processing versus timesharing, then sophisticated equipment should be the password rather than makeshift systems.

(7) There should be an understanding that about 20 percent of technical budgets will be related to computer systems in engineering practice.

(8) Because computers allow more tractable approaches to dealing with non-linear problems, engineering education must change to recognize the strength available in solving non-linear problems. Curricula should have increased emphasis on numerical analysis and application in dealing with non-linear problems.

General Conclusions & Recommendations 17

(9) Laplace transforms have had extensive application in treating engineering systems, and now Z transforms must be included in our educational processes.

(10) The Accreditation Board for Engineering and Technology should watch carefully for the need for students to understand numerical analysis better in order for them to exploit the network-joinable desk computers.

(11) Because of costs in use of software, there might be some reduction in extensive use of software. Therefore, industry should be willing to license software to universities at reasonable prices in order to stimulate the education of engineers, both in the undergraduate and graduate programs, in the use of desk computers joinable to networks.

E. Conclusion: Advisory committees consisting of industrial and governmental members can be very effective in changing and supporting engineering education.

RFA:

(1) Increase the number of state legislatures on advisory committees in order to increase the state's role in supporting the aims and goals of engineering education both in the public and private sector in a given state.

(2) Have strong interaction with industrial representatives on advisory committees in order to make clear the needs for grants, fellowships, scholarships, and useful, cooperative educational programs.

(3) Use advisory committees to increase communication on development of the highest quality of guest lecturers, faculty members on loan to industry, and industrial members on loan to faculties.

(4) Exploit advisory committees to increase communication on current technical information to the graduate and undergraduate programs in engineering education.

(5) In the establishment of advisory committees, emphasize the opportunities for high corporate officials to participate as well as other members of industry and members of governmental bureaus.

(6) Note that many problems to be considered in engineering education, and particularly in

ENGINEERING EDUCATION

case studies, can have tripartite solutions that deal with the university, government, and industry.

(7) Use advisory committees to communicate quantitative information to legislatures about values of engineering education.

F. Conclusion: Industry has a strong role in the education of engineers by way of providing unrestricted funding to universities, sponsoring specific research programs, funding equipment or space, providing adjunct professors, allowing consulting in summer opportunities for faculty, and for encouraging employee-alumni contributions.

RFA: Because the industrial role in engineering education is not always fully understood in detail, increase that understanding by way of advisory committees for engineering curricula wherein the advisory committees have strong industrial representation.

G. Conclusion: There must be continuing improvement in the value of engineers as related to cost of education and eventual productivity. That implies, for example, that employers might even look outside the United States for training of engineers as a more cost-effective avenue to provide engineers.

RFA: Universities in the United States should maintain alertness to be sure that there is full communication among industry, government, and education on how engineers are educated in the United States and what net values are in the future in the use of those engineers relative to alternative means of providing engineering knowledge for the advancement of industry and of our government.

H. Conclusion: The value of engineers to corporations and to the government will continue to increase because of the increased sophistication of the educated engineer and the increased opportunity for the engineer to apply computers and new equipment in attack of difficult systems problems.

RFA: Increased value of the engineer suggests perhaps fewer engineers per billion dollars of product, but not fewer engineers. That again emphasizes the need for examination by the university, private industry, and the government of the combined roles of engineers, engineering technologists, and engineering technicians in order to optimize the productivity of the country in its technological output.

SUMMARY AND RECOMMENDATIONS

THE PRE-COLLEGE EXPERIENCE FOR ENGINEERING EDUCATION

by

David R. Reyes-Guerra
Executive Director
Accreditation Board for Engineering and Technolgoy (ABET)

If requests for admission to engineering colleges are taken as evidence of success of the ongoing guidance, recruiting and public information efforts of the engineering profession, then it appears that these programs are very successful. However, the motivators for students to consider engineering are not necessarily based on such programs. There is evidence that the interest in engineering is in large part due to public exposure to the high demand for engineering graduates and the high salaries - as compared to other college graduates.

There is overall approval of the ongoing efforts - the only criticism is that they may not be sufficient in number to reach a larger segment of the intended audience - students, teachers and parents.

Interest in an engineering education is high. The pervasive problem however, is not interest but rather the prospective students proper pre-college preparation which would insure reasonable success in the most demanding engineering curricula without the need for extensive remedial work.

Since the current admissions situation is highly competitive, students not only need motivation but also proper academic preparation. To inculcate this need for pre-college academic preparation, it is agreed that competency and understanding of mathematics and science is a necessary element.

It is evident that such competency must be generated as early as possible in a students academic career. The high school years (9-12) should be the capstone to an interest that has been generated and cultivated in grammar (1-6) and junior high school (7-8).

It was recognized that an effort must be made to address the problem of insufficient and inadequate mathematics and science teaching in the early years.

Competency is not the only necessary element to insure admission and success. It must be coupled with a strong guidance effort to determine and reinforce interest and motivation.

To impact these identified needs, several actors were identified:

Government, Industry, Teachers, Engineering Educators and Engineering Societies, and Members of the Engineering Profession.

Government - federal, state and local - has a major responsibility in public education. The economic status of the country in no small way hinges on the technological literacy of the citizenry and the optimization of the intellectual capability of individuals.

Industry not only recognizes its obligation to the population it serves through its products and services but also its role in the community. It is an integral part of the publics concerned with education.

Engineering educators can substantially affect the pre-college educational endeavor by providing assistance, within the educational domain, to pre-college teachers.

The engineering profession through its societies and individual members recognizes its responsibility to the profession and to the public it serves.

There is support for the following conclusions:

(1) Students need exposure to math, science and communication skills, presented by competent teachers in an interesting manner, in the early years of their education.

(2) Because we live in a technological world, we need to promote technical literacy among all age levels of the population.

(3) We must recognize the need for special programs and special approaches to education for minority students until they become fully integrated (culturally, educationally).

(4) The local community is the core element that can impact education. Engineers need to be involved at the local level to influence education.

(5) Education can be enhanced through appropriate use of resources already available.

(6) Teacher education programs need to be impacted to assure competency in subject matter on the part of their graduates.

SUMMARY AND RECOMMENDATIONS

ENGINEERING EDUCATION AND THE BACCALAUREATE EXPERIENCE

by

Russell R. O'Neill
Dean of Engineering
University of California - Los Angeles

INTRODUCTION

ENROLLMENT

70% growth during past five years
Many schools and colleges have (or will) limit admissions
No discussion of steady state needs of U.S.

STUDENTS

Cream of high school crop, but crop is getting poorer

CURRICULA

Responsive but time constant is long. High mortality rate for innovation
Diverse over wide range
Four years has become standard

5 year and pre-engineering plus professional school discussed
4.7 year average (may reflect poor preparation in H.S.)

Fundamental Education (math, science, engineering science) quality high
Liberal * Education best on campus but could be better

Humanities are unfocused
Communication skills inadequate

Both will be much more important in 80's

Problems must be solved humanely and implemented thru communication

Remember that:

* Some prefer "liberating"

Engineering profession is open-ended
Curricula neither necessary nor sufficient

Things to watch for:

There is an important distinction between science and engineering

> Mode of thinking
> Applications - design - laboratory instruction
> Economics

State of the art is changing rapidly and students must be able to make the transition from the university to first job.

Several forces will have enormous impact. They are:

> Info-sphere revolution (Toffler "The Third Wave")
> Personal computer
> Genetic Engineering (biology)
> Microprocessors, artificial intelligence, robotics, (will change processes as well as control them)
> Education in other countries will impact ours
> Management of curricula i.e. planning, developing, integrating the humanities
> Supply of teachers is becoming critical
>
> > **alternate sources esp. industry**
> > large classes
>
> Cost of instruction
>
> > Engineering faculty salaries
> > Faculty frustration
> > Laboratories

RECOMMENDATIONS

Maintain Quality

> If not country's quality of life will be lowered

Maintain Quantity

> If not quality of life will be lowered
> Social gains lost, we must be sensitive to minorities both at admissions and retention

Maintain Relevance

> If not the industry will not be served. They will set up "Exxon Institute of Technology", etc.

Strengthen Humanities Contribution

    If not problems will be solved but solutions not implemented

Improve Communications Skills

    If not problem solutions will not be implemental

Obtain Additional Dollar Support

    If not there won't be enough

Use systems approach to solve problems in engineering education

SUMMARY AND RECOMMENDATIONS

GRADUATE EDUCATION IN ENGINEERING

by

Robert H. Page
Dean of Engineering
Texas A&M University

INTRODUCTION

A pleasure to participate in this conference of concerned individuals. Although these summary comments are my own, they are greatly influenced by your interactions, comments, and presentations, from which we all have learned so much during this meeting. I will --

(1) Summarize session (and hallway) comments
(2) Present Consensus View (as I see it)
(3) Present Recommendations for Follow-up
(4) Present Consequences of No Action
(5) Present Consequences of Action

SUMMARY OF SESSIONS

* Microelectronics and Computer Graphics will have major impact

* USSR is graduating 250-300,000 engineers per year at equivalent of the U.S.'s Master's level

* Engineering Education at Graduate Level in countries that successfully compete with U.S. in technological areas showed opportunity for more study

* Positions in universities which graduate students fill (e.g. fellowships, teaching assistants, research assistants, etc.) should be made more attractive

* Yearly small cuts made in quality, when compounded, have a great influence

* Plans for solutions to graduate education problems must come from universities

* New approaches may include mini-courses of 1,2,3, or 4 weeks length in order that there can be a flow of engineers between industry and the College of Engineering for short term assignments

ENGINEERING EDUCATION

* A productive track record in Engineering Design activities should be recognized as a faculty qualification alternate to a doctoral degree

* Engineering faculty at all levels should reflect in a professional manner the qualities of our free enterprise system

* At all levels, engineering faculty are desired which are role models for minority students (i.e. minority faculty)

* Reindustrialization of U.S., with increased emphasis on low and high technology, will require a greatly expanding supply of Master's and Doctoral Engineering graduates

* Past data shows only 1 out of 3 or 4 Engineering Doctors end up with employment in the university

* A 10% increase in new technology development could result in a 3 billion dollar per year increase in GNP

* A substantial number of additional fellowships and traineeships are needed with high starting salaries

* Current stipends for assistantships, etc. must be increased

* Lab facilities need modernization and new research equipment for doctoral level research secured

* Professional environment of faculty must be competitive with industry and government

* Industry should encourage best B.S. graduates to consider graduate study

* Federal role will be to support _basic_ research in future

* Industrial role will be to support _applied_ research in future

* NSF hopes to increase graduate fellowship stipends for engineering

* Graduate faculty currently find industry has a more attractive career path with equipment and support services

* Undergrads are often receiving a 4 yr. B.S. degree for a program that takes more than 4 yrs.

* Industrial support of research greatly assists graduate education (e.g. Exxon-MIT deal is most attractive)

* The concept of a Professional School of Engineering and a possible model deserve consideration

* A comprehensive study with a generalized analysis of all levels of engineering education is planned by NRC

* Crisis in engineering education focuses on Graduate Education since the <u>future</u> lack of qualified engineering faculty could result in self-destruction of the Engineering Education System

CONSENSUS VIEW

* Serious problems exist in providing a supply of Master's and Doctoral Graduates to meet the need

   * Need: Continued and Future high technology growth of U.S. and improved productivity of current U.S. industries, require a quantum jump on the production of Doctoral graduates from U.S. Engineering Colleges

RECOMMENDATIONS FOR FOLLOW-UP

* Attractive Graduate Fellowships are needed
* Laboratory and Teaching Equipment need modernization

CONSEQUENCES OF NO ACTION

* Growth of high technology in U.S. will be impaired
* Brain drain from Engineering Colleges to (?) will occur
* Nation's capacity to train B.S. level engineers will decline steadily

CONSEQUENCES OF ACTION

Bright future for U.S. in a technological world with improved U.S. standard of living, an internationally competitive country, and a strong defense posture

SUMMARY AND RECOMMENDATIONS

ENGINEERING PRACTICE AND ENGINEERING EDUCATION

by

Charles E. Schaffner
Executive Vice President
Syska & Hennessy, Inc.

My assignment, according to the program, is to present a consensus including recommendations for follow-up on Engineering Practice and Engineering Education. Obviously, a consensus is impossible, so what I shall do is to give you my point of view - crediting other people when their viewpoint coincides with mine - and including an emphasis on responsibilities of the Accreditation Board for Engineering and Technology (ABET) and the American Association of Engineering Societies (AAES) as organizations representing the entire profession.

Shortly after the formation of the AAES in January, 1980, President Carter expressed concern about the status of engineering and science education and asked for the preparation of an Administration position paper. The American Society for Engineering Education (ASEE) was asked to submit material to the Administration. As President of ASEE and Chairman of the Educational Affairs Council of AAES, I was able to ensure that the material prepared by Don Marlowe was endorsed by the engineering profession through AAES as well as representing the position of ASEE.

One of the problems we discovered in attempting to validate our position was that much of the data was either contradictory, unsupported, or fuzzy. You have been exposed to some of the data this week. One of the initiatives that industry has taken you heard Ed Lear describe on Monday. Eight corporations have pledged enough money to AAES to enable ASEE (The Educational Affairs Council Secretariat) to retain Jack Geils to work full time on the problems of Faculty, Graduate Students/Equipment/Space and in the process to document with hard data the specific needs of the institutions.

We have heard a number of views during the week on the definition of engineering including a view that it can't be defined. As a setting for the remainder of my remarks, I would like to give you mine.

e   The one basic purpose of engineering is the economic construction of a needed product. Therefore, applying the engineering method to the design of an engineering education system dictates that the purpose of engineering education should be the economic construction of an educational system designed to prepare a student to economically construct a needed product. The economic construction of such a system dictates the inclusion of a significant design experience.

This definition, stressing economic design and construction (or manufacture), leads directly to the conviction that the schools of engineering play a role in the determination of the U.S. national economic health analogous to the place of the schools of medicine in the provision of health services.

Obviously, the statistics indicate that the colleges are in crisis and if so the national economic health will deteriorate. What can be done about it?

One quoted statistic claims 2000 faculty vacancies; another 10% less faculty than 10 years ago and 70% more students. The latter combination yields a far more serious situation than the first. And when one adds the drive for more graduate students, the increasing emphasis on continuing education, the increasing enrollments of minorities and women, and the technological education of non-engineering students it would appear obvious that the load will get heavier rather than lighter.

I can only conclude that Jerry Haddad is correct: massive federal aid is unrealistic and even if available, shouldn't be accepted; massive industrial aid is unrealistic; the solution is in restructuring. Such a restructuring will require a change in mind set on the part of practitioners and particularly so on the part of educators.

We simply can't continue to do business in the same old way. Whether the shortage is 2000 faculty members or - as I suspect - two to four times that figure, it will take a generation to get back to normal. And that assumes that the proposed increased graduate stipend is adopted, industry does not then raise the ante, and all of the other negative aspects of academic life are ignored by the prospective graduate students.

How do we restructure? The profession, through its practitioners, must take a responsible role in engineering education. Historically, both educators and practitioners have been satisfied with a collegiate form of education in which the practitioners took little or no part. Engineering is the only profession with a society devoted solely to education for the profession and the fact that its membership has consisted primarily of college people is further evidence that the profession has left education primarily to the educators.

We must reverse this process. The practitioner must become more involved in the hands-on process of academic life including a role in financing and curricula development, committee activity, and student thesis advising as well as in teaching.

If the American Association of Engineering Societies can stimulate the educators to reach out to the practitioners and remind them of their obligation for the education of their successors, and stimulate the practitioners to respond, we will be starting down the road to the solution of this problem.

I don't know what the policies of all the professional societies are in this regard, but the American Society of Civil Engineers policy states -

"The American Society of Civil Engineers recognizes that an important hallmark of a profession is acceptance by the members of that profession of responsibility for the education and training of those people preparing to enter its practice.

....the American Society of Civil Engineers as a society acknowledges the necessity for devoting a substantial portion of its energy and resources to civil engineering education...it is the responsibility of individual practitioners to participate directly in both on-campus and off-campus education activities...

....a majority of the faculty of civil engineering schools should be themselves successful practitioners....."

I am proposing that the suggested restructuring take the form of other professions such as medicine, law, or architecture where a substantial part of the teaching is done by practitioners and where a substantial portion of the full-time faculty have been in engineering practice and are currently active to some degree.

I am proposing that we do away with - now and forever - that abominable term, Adjunct Professor, which has a number of connotations- all bad. Recently, Dan Drucker in referring to his perception of the full-time educator, called him a Researcher/Educator. I propose that we call the full-time practitioner a Practitioner/Educator. And I believe that there is room for an Educator/Practitioner as well.

There are three times as many practicing engineers with doctorates as there are on faculties. When one realizes that only about 50% of the current faculty time is spent in teaching, the effective ratio may be 6 to 1. If one then entertains the forbidden thought that leading practitioners who do not have a doctorate just might be able to teach effectively while also providing an excellent role model, the ratio skyrockets.

The manpower is there, it has the ability, and under the right conditions, it will participate. But just as modern management stresses participation, many of the best practitioners will want a larger role than just a teaching assignment. I am convinced that the combination of the Practitioner/**Educator**, a **functioning Industry Advisory** Committee, and Practitioners as Board members will reverse the pendulum swing in terms of faculty loads, compensation, consulting opportunities, funding and equipment.

Obviously, this can't be accomplished overnight in all school in the country. But it can be started in all to some degree. And it would be interesting to see one or more institutions embark on a determined effort to institute such a reorganization in a relatively short period, as Worcester Polytechnic Institute completely changed its educational philosophy recently. Perhaps NSF or the Sloan Foundation might help to interest someone in trying it.

This proposal may sound radical in terms of the operation of the average engineering school of today but many of its components have been suggested during this week and some are being practiced already. And, as pointed out previously, it represents the normal practice in many other professional schools.

Now, let me turn to some other recommendations for industry and practitioners, some of which can be effectuated by their representative bodies AAES and ABET. Many of these you will recognize as suggestions by Messrs. Haddad, Kezios, Samson and Sprow. I have also drawn on the ASCE Interim Report to the Board of Direction on Engineering Education, a paper by Eric Walker, and Congressional testimony by Robert Frosch, President of AAES.

First of all, ABET in its accreditation should insist on:

Competitive faculty salaries
Realistic faculty - student ratios
Appropriate allocation of institutional resources
The establishment and effective use of Advisory Councils

Evidence of a significant involvement of a majority of the faculty in non-academic engineering consulting and other up-to-date engineering experience gained through continuing practice, research, professional and sabbatical leaves, and other work experience.

Faculty members having responsibility for teaching engineering synthesis and design courses be registered engineers in the state in which the educational institution is located.

Evidence of commitment of members of the profession from off-campus in the educational process including establishing educational objectives and if, as suggested this week, some states or institutions wish to withdraw from the system - so be it. Let us find out whether we are - in fact - a profession.

AAES through its Educational Affairs Council should:

Encourage member societies to designate at least one of its two delegates as a practitioner.

Encourage educators and practitioners to develop the institutional organization recommended earlier.

Take the leadership in the overall guidance effort and particularly in the coordination of the efforts of the organizations active in the encouragement and development of minority high school students including NACME.

Take the leadership role in the education of high school and non-engineering college students, the media, and the political leadership on the role of technology in our society coordinating with the Council for the Understanding of Technology in Human Affairs.

In its work in student development, encourage the activities of the minority organizations such as NSBE, SHE, MAES, and the AISES.

Encourage engineering educators to increase productivity through efficient management of learning by maximum utilization of teaching technology.

Encourage engineering educators to utilize peer teaching to prepare students for the role as teacher of subordinates, clients, and the lay public.

Encourage engineering educators to utilize cooperative (as opposed to individual) learning experiences to prepare students for the actual practice of engineering which is essentially a team effort.

Other general recommendations, many of which are already being done someplace are:

    Utilize faculty on sabbatical leave.
    Utilize faculty in consulting arrangements.
    Lend key people for short time periods.
    Donate, or sell at heavy discount, current state-of-the-art equipment.
    Provide opportunities for utilization of industry equipment on site.
    Provide input on curriculum development.
    Increase the use of cooperative students.
    Increase summer job opportunities, particularly for minority students.
    Provide leaders for graduate thesis committees.
    Establish specific industry - institutional alliances. There is no reason for every engineering college to be everything to all industries and to all students.
    Encourage government to provide incentives and to catalyze arrangements among universities, government, and industry that would provide industry support.
    Encourage NSF to provide catalytic support of research as a cooperative effort between industries seeking innovative help and university graduate research and education which can assist in providing it.
    Encourage NSF to support research in areas not covered by the mission agencies.
    Provide facilities on industrial site for teaching.
    In coordination with AAES on national level and institutions on local level, lobby for increased government support of engineering education.
    Sponsor academic research.
    Provide unrestricted grants.
    Encourage employee-alumni gifts.

During this week we have attempted to influence positions somewhat like the British parliamentarians who once demanded of a British Prime Minister of yore "Are you going to grant home rule to the Irish?" "I will," he replied, and after the applause of the proponents of such a measure had died down, he continued, "not". This brought ovations from the opposing faction. When silence reigned once more, he completed his statement with the word "say"; this time bringing down the house.

I know that Bill and Lee and the committee will say, and hopefully what is said will be heard, and acted upon, by institutions, government, industry and the practitioners.

SUMMARY AND RECOMMENDATIONS

RESEARCH AND ENGINEERING EDUCATION

by

Richard G. Cunningham
Vice President, Research & Graduate Studies
Pennsylvania State University

Baccalaureate programs are the primary contribution to the engineering profession and to the country. Graduate programs and research play key roles through contributions to knowledge, research manpower training, faculty development, and enhancement of teaching.

One problem stands out: The serious erosion of U.S. student pursuit of advanced degrees in the face of high demand and salaries for baccalaureate graduates. We are approaching a literal cut-off in supply of our future faculty and advanced design and research engineers for industry and government.

Three sets of recommendations have been identified, as follows:

To the Federal Government

1. A marked increase in industry-university cooperation in research and development is both desirable and feasible. Tax incentives to encourage industries to purchase basic and applied research from universities are strongly recommended.

2. The shortages of U.S. students in graduate programs, are so acute that direct stimulus is needed in engineering and in computer science and engineering. Fellowships or traineeships - with stipend levels set reasonably to compete with the high salaries paid to engineering baccalaureate graduates - are needed, without further delay. At stake: national productivity and national defense manpower.

3. Laboratory and computer equipment costs are unsolvable problems to universities. Inflation, obsolescence in a five-year time span, and high maintenance costs for sophisticated devices all add up to unprecedented demands on school budgets. Federal assistance in grants and indirect-costs allowances are needed, in the interest of raising or at least maintaining quality of engineering education.

ENGINEERING EDUCATION

To Industry

1. University faculties and facilities are a valuable national resource in improving the U.S. posture in R&D. Increased purchase of basic and applied research in academe is recommended. Universities can and will become more flexible in accommodating needs of industry, than in the past when Federally sponsored research predominated.

2. Assist engineering schools in attracting top students to graduate study via summer or co-op jobs and delayed employment arrangements. Recognize that increased minority participation in engineering - in industry and academe - depends on graduate programs.

3. Most companies visit campus to recruit, only. Participate in and help your MP supply line. Encourage your engineers both young, and upper-middle management levels, to serve on advisory boards, accreditation team visits, design project evaluations, and to do so on a long-range continuing basis. Industry can meet a national obligation and at the same time benefit in terms of better and more stable manpower supply.

4. Consider a formula-based direct assist to your sources of engineering graduates through grants in proportion to number of hires.

To Colleges and Universities

1. Administration must recognize that industry is academic competition for the intellectual capital essential to quality engineering degree programs. Industrial starting and continuing salaries should be identified, and met, or at least approximated.

2. Industrial competition similarly must be met to ensure continuation of graduate programs. In the face of $25K starting salaries, assistantships at $4-$6K are out of touch with reality.

3. Faculty can and should sell graduate study opportunities to the top 10 or 20% of undergraduates, and universities should encourage faculty action to strengthen enrollments.

4. Compared with the exciting and well-paid opportunities in industry, academic careers are no longer very attractive to young graduates. Low pay is coupled with the gamble of pursuing combined teaching-research duties down to the 7 year tenure-or-not deadline. New approaches to hiring and retaining engineering and computer science faculty are recommended.

5. Faculty should be encouraged and helped in establishing contacts in industry leading to research contracts.

University policies will require modification. Flexible patent policy, and greater tolerance for industrial sponsor screening of proposed publications for proprietary information are two examples.

6. Replacement of some Federally supported research with industry sponsored applied research can have a healthy effect on the quality of engineering education by providing better balance.

7. Laboratory and computer equipment problems require serious study and new approaches. One-shot infusions of Federal money - even if forthcoming - will not solve this ongoing deficiency in engineering education at all levels. Debt financing for equipment purchasing should be considered seriously.

8. Some of the new high-technology areas will be so equipment and capital-intensive that most schools may not be able to pay the costs. Selectivity and long-range planning will be essential. (Examples: VLSI and genetic engineering).

9. Traditional semester-long sessions make life difficult for top ranked engineers in industry, interested in graduate study or in teaching. Modules of 1-4 weeks will attract greater industry participation.

10. Graduate programs in engineering probably have copied science too closely. Education in Japan and Europe should be evaluated. One example of a change to be considered: A higher earned doctorate, similar to the Doctor of Science in England, the Doktor Habil in West Germany and the Doctor of Engineering Science in the USSR are three examples. Benefits would include a stimulation to faculty and engineers in industry to make fundamental contributions to engineering science and design; and better use of sabbatical leave.

OPENING REMARKS
July 27, 1981
LELAND J. WALKER, P.E.
PRESIDENT, ABET
CHAIRMAN, BOARD OF DIRECTORS
NORTHERN TESTING LABORATORIES, INC.
P.O. BOX 951
GREAT FALLS, MONTANA 59403

Now that you are all here and ready to begin, you may wonder why ABET called this meeting!

The theme of this conference - "Engineering Education - Aims and Goals for the Eighties"--would stand alone under normal circumstances and the relatively smoothly functioning system that we have enjoyed generally for the past twenty years or so.

However, the circumstances are not normal, and we are beginning to see our smoothly functioning system encounter some serious bumps, obstacles, and problems.

In my business, when something like that happens, there is a clear understanding of who is responsible for action, and a rather clear-cut mechanism for developing a response to the problem.

Such is not the case in the engineering education system. That system is based on an intricate inter-relationship between the educational institutions, the engineering profession, industry, and government. This inter-relationship is fragile, in that it is mostly ad-hoc and voluntary, and no one is really "in charge." I hasten to add that in my judgement, this has been an important strength of the system in that no one point of view has been predominant, and our mutual respect, cooperation, and understanding have provided well for engineering education under normal circumstances. But circumstances are not normal:

*   Engineering enrollments - In five years undergraduate enrollments have risen by 70 per cent, from about 200,000 to 340,000.

*   Engineering schools are beginning to limit enrollment in several ways - by placing a cap on total enrollment, by raising entrance requirements, and by raising continuation standards.

*   Recruitment - In spite of those figures, corporate recruiters have found increasing difficulty in filling new and vacant positions.

*   Financial support has not kept pace with inflation, let alone with the increase in numbers of students being served. Support from the Federal Government (including NSF) has decreased, and may well

decrease even more. State support has decreased, as a result of
legislation and the so-called taxpayer's revolt (i.e., Proposition
13, Proposition 2½, etc.)

* Laboratory facilities and equipment have become obsolescent, as
well as being inadequate for adequate student instruction.

* Faculty are aging, are being attracted by opportunities in
industry and in consulting, and the normal source of new faculty-
from graduate programs - is shrinking rapidly.

That's why the meeting was called!

But why ABET?

While ABET is only one part of the "system" it has a unique role
in that it has input and involvement from education, industry and
practice, but at the same time is independent, as the quality control
standard-setter and evaluator for engineering education. As such, it
can be perceived as not self-serving in its objectives. I should note,
however, that developing solutions to the problems facing engineering
education is self-serving, in that if we are successful in finding
solutions, our quality control job will be much less difficult.

The educational institution "report card" for last fall's round of
visits has deteriorated significantly from previous years -- it is
never easy to give C's, D's, and F's when the norm is generally A's,
B's, and C's!

In developing the agenda for this conference, we are indebted to
a large, distinguished steering committee, many of whom are here with us.
We appreciate the support of the National Academy of Engineering, the
American Association of Engineering Societies, and the American Society
for Engineering Education, who are co-sponsors. We are particularly
grateful to the Engineering Foundation and its Director, Sandy Cole,
whose sponsorship has made the conference possible and to David R.
Reyes-Guerra, our Executive Director, who suggested the meeting and
joins me as co-chairman. Finally, I want to introduce Dr. Bill Corcoran,
Vice President of ABET, who is the Chairman of the Program and Steering
Committee, who planned and organized it all, and who will now tell you
about the format of the conference, and what we are committed to achieve
by noon on Friday.

AN EXAMINATION AND DISCUSSION OF OPPORTUNITIES
AVAILABLE TO GIFTED STUDENTS WITH ENGINEERING
INTERESTS
PROBLEMS AND INTERPRETATIONS

by

Lee F. Browne
Lecturer/Director of Secondary School Relations
and Special Student Projects
CALTECH
210-40
Pasadena, California 91125

Extreme giftedness, to some degree, has historically been suspect in America.

The condition is generally looked upon as being possessed by people who practice excessive bookishness and an idealism which borders, in many people's minds, on freakishness.

The general attitude (historically) on the part of the average American was that we should maintain a generalized approach toward acquiring intelligence; that is, we should glorify mediocrity.

It emerged that by the forties and early fifties, we had embarked on something which was to become known as "do your own thing".

Emotionally, it was generally felt that if you (as a gifted person) did things too well, or if you spoke too articulately, or if you understood science too well, you were what was generally called an "egghead"; and in many quarters it was observed that you must be hell-bent on changing America.

You might even be a communist.

Many felt that it had been "demonstrated" that idea-type people were not in tune with free enterprise or business, but with ideas - not generally oriented toward making money, but making "intellectual trouble." Some thought that people like this were too smart to be trustworthy, especially if you were an economically disadvantaged Caucasian, or a non-white.

If you belonged to this latter category, you could survive if you were a "tinkerer" or if you kept your ideas to yourself, or if you paid general public and overt homage to mother/flag/country and GOD. In many discussion areas, objectively on your part might negatively label you.

If you were a non-science/math educator who sought to collect all (or as much as possible) of the data, and present this elaborate data to students you were suspect; and might be called on the carpet or even have to take "an oath".

Many such educators were subject to this oath-taking phenomenon during the McCarthy period.

Math and science educators were not likely to be subject to this type of problem because the subject matter that they taught was rather precise and clear-cut.

In the basic courses of mathematics and science, and to some extent engineering, at both the secondary and the baccalaureate level, a teacher rarely presented information which had a controversial nature. They might present two or three different models of a concept, or talk about the historical development of a theory or concept, but only occasionally would there creep into the discussion matters which were related to sociology.

At the high school level, there were a few exceptions in matters related to birth control and human physiology. And during the ecology years, some controversial environmental discussions. But, by-and-large, math/science and engineering classes, professors and students appeared to be what one might call conservative.

During the whole free-speech movement, at the University of California, during the time that people were clustering in the quads to discuss various and sundry controversial matters, the math, science and engineering people were absent from those discussions and were found busily engaged in their activities in the classrooms and laboratories. This was also true during the student Viet Nam peace rallies.

This, in some way, accounts for the general lack of understanding of math, science and engineering types by other students and by other staff. The other disciplines are constantly engaged in rather serious and in-depth examination of diverse ideas. And, generally speaking, scientists who became involved in sociological-type activities were more-or-less frowned upon for a number of reasons. Some good examples might be, for an instance; Linus Pauling who, in his activities as a chemist, also participated fairly heavily in peace-type activities and, consequently, was the winner of a Nobel Prize for Peace.

This was not graciously looked upon by other scientists because many of them felt that he should restrict his activities to science.

Even Dr. Shockley, the Nobel Prize winner for the development of some type of electronic vacuum tube, who has been used and quoted variously on matters relating to the relative level of intelligence as demonstrated in various racial groups, is looked upon by some scientists as being rather controversial.

But, by-and-large, science students, mathematics students, engineering students lead a rather cloistered life, to some extent monastic, and do not overlap greatly in campus activities; and are looked upon by others in the academic community as too quality-oriented, rigid and to some extent anti-people.

This phenomenon of isolation on the part of math/science and engineering students and faculty, to some extent, has worked against having larger numbers of so-called gifted students enter these professions. This is not to say that most of the people who go into science, math and engineering are not highly intelligent, but that many more gifted students would consider these options if there was room for exploration, or the interaction of ideas with experimentation; or the adaptation or application of some of these experimental concepts and ideas to the human condition in a sociological way.

If the pursuit of knowledge in these areas were not so rigidly restricted by curriculum, more students might be involved. When one picks up a college catalogue and looks at the manner in which one fulfills a degree requirement for a Master's in applied science, it's extremely specific in terms of the courses which must be completed. Whereas, in English and some of the other social science areas, there is a level of overlapping between courses and electives. Time is available so that the student can broaden his background and overlap into other disciplines.

This level of discussion should not be confused and interpreted as desire to water-down the curriculum of math/science and engineering.

There is obviously no way that this can be done.

However, it is possible to extend the course requirements into some of the humanistic areas to a greater degree.

This might mean the development of a professional school of engineering. A school where there is a central core called the general studies division, and in that core division, courses in anthropology, social science, art, the classics, and music, which would be seriously taught by scholars with as much scientific and mathematical infusion as possible. Clustered about this general division would be the typical schools of engineering: mechanical engineering, electrical engineering, chemical engineering, environmental engineering, nuclear engineering, and a new interdisciplinary school of engineering.

We believe that this approach would offer greater opportunities for gifted students to participate. The hypothesis is: A gifted student develops or approaches learning quite differently than do students of specialization. Gifted-type students tend to want to explore the core courses and the idea-type courses and the rigid academic courses; and they want to do this in such a way that these courses overlap. They do not, it appears in their early development, wish to study specialized courses as much as interest-oriented-type students. For this reason, we feel that students of this type would profit from going to a professional school of engineering which was previously described. Then these students would find it very interesting to do engineering research, or to develop think-tank-type activities which would give them the opportunity to develop new engineering processes in these various options.

This would particularly be true after they had had this rather generalized background, and then become specialists.

The educative process may take a little longer for these students, but in the long run, it would produce the kind of qualitative leaders in the engineering sciences which would put us back into very serious competiton with West Germany and Japan and a couple of other countries. And as time passes, it would tend to project us far beyond anything which is now known.

This is not an indictment of American education; it is simply a rather objective view by a person who has visited many engineering schools, many high schools, many colleges, and has met a number of students who have very high SAT's and Achievement Test scores, and high IQ scores.

We have watched some of those students get involved in rigid academic programs and become bored and subsequently drop out.

We believe that the Henry Fords, and the Thomas Edisons, and the George Washington Carvers, and the other technological-type engineers and scientists which have been accidentally produced in this country can be educationally produced.

It might be suggested then that education in the **eighties might** address the problem of scientific and scientific-application-type education in such a way that we revamp the science/math education programs in our schools, starting in the kindergarten and going through to the University.

The goal would be to produce more creative people who have a knowledge of mathematics and science.

We believe that there is an opportunity for gifted students in this area, but we believe that the manner in which these students are educated early will determine the number that are attracted to math/science and engineering.

There are some interesting specific historical examples of America's reaction to "eggheadedness". If you remember when Adlai E. Stevenson ran for the Presidency, it was occasionally thought by some that he was too bright to be the President; that he was too full of fuzzy ideas. He might, in some ways, be connected with that segment of the world which tended to give capitalism, perhaps, another slant.

Then, all of a sudden, there was a small metallic object which was shot out of the Soviet Union and began to circle the globe. The object was known as Sputnik. Under the Sputnik cloud, the quest toward intellectualism in the United States, especially scientism and an understanding of mathematics and problem-solving, flourished. **From about** 1958 through 1964, there was a tremendous amount of money spent in secondary school curriculum development.

There emerged the BSCS Biology Series, the blue, green and yellow versions which were cooperatively developed by the American Institute of

Biological Science and the National Science Foundation.

There emerged the Physical Science Curriculum Study which took place at MIT. It was to become known as PSCS.

And then there was the million-dollar project which was developed under the National Science Foundation which subsequently became known as Chem-Study.

All of these curriculum studies which beefed up secondary school science in the United States were coupled with annual summer institutes and, later, the academic year institutes.

Many, many secondary science/math teachers went to these sessions and acquired updated subject matter in Biology, Chemistry, Physics and Math. And it began to show.

The United States became very successful with its satellite program, and by 1964 our satellite program tended to quelch this trend toward scientism, or decrease scientism in the United States.

Our man on the moon series were, in fact, so successful that everyone assumed that we had it made.

And after the middle-sixties, science and math aspiration declined. We reentered the open curriculum.

Let me give you an example: Between 1959 and 1964, I worked at a public high school in Pasadena. And in 1959, there were 15 chemistry classes, 6 physics classes, 2 calculus classes, and about 6 or 7 trig classes. This was a **public high school**! which was not in an entirely middle-class area - it was a rather comprehensive area in terms of socio-economic and racial composition.

The so-called "better high school" on the east side of town, during that period, had comparable classes. By 1964, there were 7 of these chemistry classes, 4 trig classes, 1 calculus class which was half-full, and 2 or 3 physics classes - a reduction of almost 50%.

Out of this emerged something which was subsequently going to be called "back to the basics". This concept was very controversial in many intellectual areas in the United States.

It appeared to have several political and sociological overtones. I am going to cite these overtones as being kind of a collective compendium of what some educators thought but would not state publicly.

First of all, the "back to basics" concept had several general points of view: One of them was to make sure that people could read and write and do arithmetic, and fill out forms, and generally understand the fundamentals of selected subject matter.

It became known as something like fundamentalism, essentualism.

It was designed to, according to some people, "take care of the unpatriotic and undisciplined segment of the population in the schools."

This, in many ways, was possibly directed toward certain minority or marginal groups. Therefore, in its entirety, in tis simplest terms, it appeared to have an anti-minority, anti-marginal student orientation. The approach was to "clean up" this segment of the school population, and transform them into useful consumer-oriented Americans who could read and write and spell and do numbers, and who were patriotic, and who were going into business.

Others have suggested that it was designed to show how illiterate in math and science and reading and writing and behavior some persons were. And this would make other segments of the school population see how clean-cut and important they were.

I don't particularly buy this. But the persons who were engaged in developing these "back to basics" concepts did show some tendency toward being anti-democratic.

The really unfortunate thing about the "back to basics" movement was that it did take hold.

Unfortunately, the "back to basics" concept became the curriculum. In other words, teachers did not just teach the "back to basics" and then move to the next step, and then move on the the next step - but it became the total curriculum.

This stifled and watered down many strong programs at the high schools. This preoccupation with "back to basics" - with its love of motherhood, the flag, patriotism - became the thing, rather than trying to develop what had been called "eggheadedness".

This technique made many so-called gifted students and teachers gloat over how intelligent, productive, creative and well-behaved they were; because they could master the subject matter in this quasi-intellectual environment which was, in many cases, taught by status-quo-oriented-type teachers who, some have said, were trying to capture - or recapture - the "good ole days".

The next outgrowth of the "back to basics" movement was the development of the so-called proficiency examination in many states in the United States. If one has any doubts about what has been previously stated, it becomes fairly obvious that these proficiency exams were directed toward certain segments of the population. As evidenced by the people who could not pass those tests because they did not have the basic skills.

The other persons in the population could "pass" those tests, which tended to prove a point of view - that there was this one segment in the schools that was really very good, and another segment which was really very bad.

The unfortunate thing was that those proficiency tests were not in any way designed to bolster gifted-type activities in the schools, only the most marginal kind of mediocre activities.

There was further evidence that this kind of "back to the basics" approach was catching on by the general decline of the SAT Tests and the Achievement Tests - and the AP Test scores. Some educators feel that the decline of SAT and other test scores are directly related to the national emergence of the "back to basics" programs. The educators who are developing these SAT-type tests do not teach in the "back to basics" programs and consider such programs to anti-intellectual.

There was also a reduction in math and science classes as was previously stated.

There was a <u>serious</u> drop and decline in problem-solving skills. But the "back to basics" people had won and had done what it was they were trying to do, and everybody was happy.

And then there was the emergence of West Germany-type quality.

And the emergence of Japanese perfection and their use of robotics.

And it was all very interesting because we used to ridicule Japanese products; now all of a sudden, here was a group of people who had not stopped with "back to basics", but who had taught the basics, then the next step, then the next step, then the next step, and had taught what we generally call "gifted" kinds of programs to large segments of their population.

While we were not only watering-down courses at the secondary level, but practicing grade inflation to play some type of academic game:

Everyone could go to college....even if you go the the JC first... 8% could transfer...it is better to have been at college one term than not at all...with two years of math you can go to engineering school - you can catch up or transfer to another option!!!....

And the game continues. Gifted students must be brought back into math/science/engineering and pre-med at a level of challenge which motivates and graduates them.

SUMMARY

Gifted students at both the high school and college level tend to become bored and lose much enthusiasm for <u>typical learning</u>.

As the motivation level drops, so does giftedness...if one assumes that this quality has to be used and be cultivated.

As the giftedness level goes unchallenged, the creativity somponent tends to go undeveloped.

There are many opportunities for gifted students with Engineering interests, but there should be many more.

There is, in my opinion, a basic need for developing the professional engineering school concept - along the lines of the diagram.

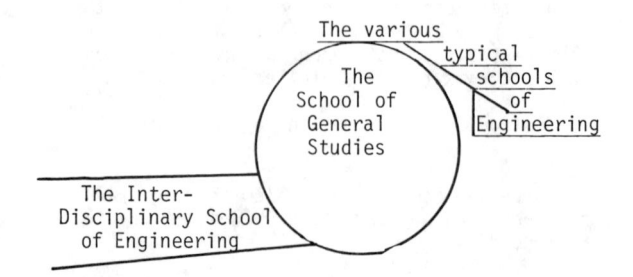

```
                   The various
                              \typical
                    The        schools
                  School of     \of
                  General      Engineering
                  Studies
   The Inter-
Disciplinary School
  of Engineering
```

This would allow more Engineering students to learn much more about each of the fields of Engineering while at the same time strengthening their <u>basic intellectual</u> skills, and developing a type of integrated understanding of their fundamental options.

Gifted students seem to be less pre-occupied with excessive quality, but more with exploration in their developmental years and, therefore, they need to be exposed to more ideas in a qualitatively rich, less rigid, but exploratory way. And as they develop, they will select specific areas for work.

There appears to be too much identification with Electronics Engineering, or Mechanical Engineering, or Chemical Engineering, among selected engineers, and not a large enough group who can successfully overlap with each other in the various options and with math-based scientists.

This direction, along with the plethora of opportunities in the typical options, may interest more gifted students to concentrate on engineering, and particularly <u>engineering research</u>.

Contrasted with "on hand" experiences and options, the greatest opportunity we can see for gifted students is in the areas of engineering research (industrial or university), and think tank type activities directed toward developing new interdisciplinary processes.

This approach would also give the engineering profession an opportunity to further democratize itself by providing equal access to these programs to indigenous minorities and women.

Some dualisms touched on in this discussion:

* Typical Learning vs Exploratory Learning

* Quality vs Creativity

* Hands-on vs Research

* Overlapping vs Specialization
* Basic Intellectual Skills vs Specialized Skills

ENGINEERING EDUCATION AND THE BACCALAUREATE EXPERIENCE

(Engineering Education - The Liberal Education of the 80's)

by
O. Allan Gianniny
Associate Professor of Humanities
College of Engineering
University of Virginia

The history of engineering education was reviewed tracing the elements and philosophy of liberal education in scientific and engineering education.

Engineering education is a liberal higher education for the way of life for this time.

Engineering is the culture of society today and permits one to best live and exist in this culture.

The following recommendations were supplied to me by Al Gianniny.

1. ASEE's REETS report should be used as a resource in organizing humanities and social science portions of the curriculum so faculty and students can respect and support these studies.

2. Schools should be encouraged to re-examine their own practices related to humanities and social sciences to reflect the School's view of their role in society, guiding students to meaningful choices.

3. Schools should include in their objectives for humanities and social sciences the relations between professional, social, cultural, and general human concerns.

4. Schools should be expected to commit people and effort to managing the humanities and social sciences portion of the curriculum, with as much care as they use in managing core areas of study and the related mathematics and science areas.

EFFECT OF PROGRESS IN MICROELECTRONICS IN EDUCATION,
IN ELECTRICAL ENGINEERING ESPECIALLY, AND
ENGINEERING EDUCATION IN GENERAL

L. T. Rader
Professor, Graduate School of Business Administration
University of Virginia

Quoting Robert Noyce, founder and Vice Chairman of Intel Corporation, "It is not an exaggeration to say that most of the technological achievements of the past decade have depended on microelectronics." (Scientific American September 1977, page 63)

First a word of definition. Microelectronics is the name applied to recent developments in which large numbers of electronic elements are put on silicon chips. These resulting circuits are called large scale integrated circuits, or LSI when the components number is in the thousands, and when the numbers are much greater they are called VLSI or bery large scale integrated circuits.

A microprocessor is a computer central processing unit (CPU) on a single chip, perhaps 0.2 inches on a side. By adding other chips to provide timing, memories, and interfaces one can assemble a complete computer system on an area of, say, 8½ x 11 inches. Such an assembly is called a microcomputer in which the microprocessor serves as the master component. (Scientific American September 1977, page 146)

This subject is the only one of our five day conference which addresses a specific technical area. The organizers of this conference evidently felt that the subject of microelectronics was important enough that it should be discussed specifically in its impact on engineering education. I agree with that assessment.

Several other topics which have been, or will be covered here, are related to, or can be related to our subject. They are the sessions one and three regarding high school curricula; session nine on high student - faculty ratios; session eleven on foreign engineering education; session 19 on the effect of computer graphics on both industry and engineering education; and session 21 on desk computers joinable to networks.

WHY IS THIS SUBJECT SO IMPORTANT?

Uppermost in the minds of many of our citizens and legislators today are major questions concerning foreign competition, national productivity, innovation and technological leadership (or lack of it), energy conservation, pollution control and more recently artificial intelligence and robotics. Microelectronics can be shown to be a potentially valuable parameter in each of these areas. In fact the Office of Technology Assessment has as one of their priority areas the subject "The Impact of Mircroprocessing on Society" and plans to do a study on this subject htis fall.

The importance of this technology is best exemplified by the fantastic growth rate it has had. The transistor effect was discovered in 1947 some 34 years ago. The first integrated circuit was made in 1958, eleven years after the transistor and the first computer-on-a-chip was invented and fabricated by M. E. Hoff, Jr. of Intel in 1969.

The chip had on it 2,250 transistors. It was called a microprocessor because it contained all the elements of a computer; logic, memory, input and output. I still remember the tremendous excitment it created in our Electrical Engineering department and we received our first chip in 1971. This was the Intel 4004 followed in 1972 by the 8008. In 1976 Intel sold a microprocessor called the 8080 and Motorola announced its 6800. Sales in 1975 were only $50 million dollars but estimated at that time (1975) to increase ten times by 1980. (Fortune February 11, 1980)

Every year since the I. C. industry began in 1960, the number of components per chip of silicon has about doubled over that of the previous year so that now 150,000 components can be fabricated and interconnected on a single chip. This growth was maintained because of the use of computer-aided design, CAD. Without the advanced analysis and extensive simulation techniques available through CAD it would have been totally impractical to design VLSI chips without taking months or years to do so.

COMMISSION ON TECHNOLOGY AUTOMATION AND ECONOMIC PROGRESS

Because of all the furor over automation and unemployment in 1964 Congress established a Commission on Technology Automation and Economic Progress. Their six volume report is still relevant and excellent but I want to lift just one quotation from it. The report said that when a significant factor can be made to change by ten to one the results are not only major, but many of them cannot be foreseen. They gave as an example that when man walked at four miles per hour it was one kind of world, when the auto arrived and he traveled at 40 miles per hour it was a different kind of world, and when the aeroplane came at 400 miles per hour the effects were still more startling and unexpected.

Using this ten to one as a general criterion, what do we see as we look at microelectronics? We see many orders of magnitude change in just a few years in size, weight, reliability, power consumption, and cost.

You have heard the figures many times and they are hard to keep straight. The major change has been in cost. The first cost of a large computer has decreased by a factor of 1000 and over the last twenty years. The minicomputer reduced it further, perhaps by five to one, for certain applications and now the microcomputer on single chips, is reducing cost another 100 times. (J. S. Mayo, Bell Labs, 1/4/80, AAAS) The result is the many prodcuts we have seen mushrooming, most of which are in the consumer field, especially the hand-held calculator, the digital watch, speak and spell games, and many TV games.

Then there are applications less obvious but already quite pervasive such as the point-of-sale recorder in the supermarket, and a large number of analytical instruments.

In general, however, the tremendous technological progress now going forward so rapidly in the fields of computers and telecommunications is due to two factors--deregulation and the large scale integrated circuit. These encompass products and areas such as:

- intelligent terminals
- word processors
- distributed processing
- electronic filing
- satellite communications
- electronic switchboards
- electronic mail and audio mail
- voice actuated devices, and
- teleconferencing

In 1979 United Technology Corporation bought a semi conductor company named Mostek. United is a conglomerate including Carrier, Essex, Otis, Ambac, Pratt and Whitney, and Sikorsky. Why did they buy Mostek? They said, "Every piece of equipment we manufacture is going to be controlled by microprocessors." (Fortune September 22, 1980)

NAE TASK FORCE, 1980

A task force on Engineering Education of the NAE issued its report in April 1980 entitled "Issues in Engineering Education--A Framework for Analysis." Some of their comments which read on the subject we are discussing follows.

Under Objectives of Engineering Education

- Of particular concern at present is declining U.S. productivity. Thus an assessment of the status of university training for industrial innovation and design is urgently required. (page 10)

- The present <u>accreditation</u> process is serving the profession well in ensuring relability of engineering programs, but little is known about the influence of the process of curricula, their quality and flexibility. (page 11)

Under <u>Resources Required to Meet National Needs</u>

- The problem has ben exacerbated by the acceleration of <u>technological progress</u> during the last twenty years, increases in the sophistication of the laboratory equipment required and increases in cost. (page 14)

- Unless the trends change, engineering colleges will not be able to provide adequate training in many of the new, most important technologies withour substantial help. For example, integrated circuit electronics requires equipment which is out of reach of most engineering colleges as do the new design methods based on computer graphics. (page 14)

    > Ohio State University recently estimated the cost of installing an adequate computer graphics system to teach modern design at $3 million plus 15 percent per year for maintenance. R.P.I. made a similar analysis and obtained similar results. The computer graphics problem is just one of many but it is an important indication. (page 15)

- A program is needed to update current engineering teaching and research facilities and laboratory equipment. (page 16)

Under <u>University/Industry/Government Relations</u>

- (A major problem is) the inability of the academic system alone to cope with increasing costs. (page 17)

- (There is a) crucial interdependence of the academic industrial sectors. (page 17)

- Strengthening university/industry ties would benefit both high technology fields and the more traditional fields. (page 18)

- At some universities multi-company support of research have been developed.

    > Silicon Structures Project at Cal. Tech. has six companies which contribute $100,000 each per year and an engineer per company for one year. (page 20) CDC and Purdue have a joint program for research in CAD-CAM--a multi-year committment. (page 20) Carnegie Mellon and DEC have a joint program. (page 21)

- A fundamental conclusion of the NAE task force is that great benefits can be realized through the fostering of existing and new ties between universities and industry. The Federal Government is in a position to enhance and encourage such interactions. (page 21)

- There are models of successful, mutually sponsored programs. (page 21)

- The ability of the universities to contribute to a systematic increase in productivity or to industrial innovation probably can be enhanced by some experimentation along this line, and is worthy of an investment in time and funds. (page 21)

- Some government programs which could be made more effective include:

    U.S. Air Force ICAM (Integrated CAM)
    DOD VHSIC (Very High Speed IC's)
    DOE (Department of Energy Programs)
    DOT (Department of Transportation Programs)
    DOC (Department of Commerce Programs)
    NASA
    DOD IRAD (Independent Research and Development)   (page 22)

IEEE Spectrum, November 1980

Under the title "VLSI goes to School" in this article are the following statements:

- Microelectronic circuits, the heart of the electronics industry may well be $10 billion in sales by 1990 or 10 percent of the value of all electrical equipment sold.

- Major programs in VLSI are underway at Cal. Tech., Stanford, MIT, Cornell, University of California at Berkeley, and

    Carnegie Mellon. It is also a major topic of research at Universities of Utah, Florida, Illinois, Rochester, Southern California, Washington University (St. Louis), and others.

SPECIAL SESSION OF THE OECD PARIS, NOVEMBER 27, 1979

A series of papers by participants from many countries were presented on the subject "Analytical Summary and Perspectives on the Impact of Microelectronics on Productivity and Employment." Some of the points made were the following:

- All the papers stressed the need for changes in the educational system and training and retraining programs.

ENGINEERING EDUCATION

- The impact of microelectronics is heightened by the convergence of other technological development which interact with it such as lasers, optical fibers, printing and display technologies, software, automation telecommunications and so forth. (page 3)

- The development of microelectronics technology presents a number of unique characteristics which have virtually no precedent in the history of technology. Its field of application is _very_ wide. There has been less time lag between inventions and commercial innovations than in earlier technologies. (page 2)

- Over half the changes in production methods in Canada over the past five years have been computer related. (page 13)

- It is also important that the curricula of universities and technical schools be restructured so that students are educated and trained for the kinds of positions that will be available in the years ahead. (page 22)

- Computer assisted instruction has been proved effective in basic training courses. (page 23)

CHRONICLE OF HIGHER EDUCATION, OCTOBER 22, 1979

In hearings before the House Subcommittee on Science Research and Technology the following points were made:

- Mr. Branscomb, Chief Scientist of IBM, said that in the U.S. computer-assisted instruction has been adopted more readily by industry than by educational institutions.

- In IBM alone, 10,000 employees of the engineering division spend 90,000 student days per year taking computer-aided courses at their offices. The expense is easily justified.

- More than a million college students have no access to computers. Said F. J. Rutherford, the N.S.F. assistant director for science education, "We do not yet know how to integrate these technologies into the classroom, or into the lives of students in a way that is effective and satisfying."

- It was proposed that there be an investigation of the status and effectiveness of _existing_ computer-based techniques of instruction and study of costs and benefits in the use of materials for computer-based instruction.

ENGINEERING EDUCATION

The following are advanced as possible initiatives to be discussed.

1. It appears that prestigious universities, both state supported or private, are actively involved in research in LSI. It is probable that lesser known shcools may not be either teaching the technology nor utilizing it. I believe therefore that State Governments should take the initiative (as North Carolina has done) to see that all their state supported engineering institutions are up to some minimum participation in this field.

2. Rather than rotation of porfessors between industry and academia, it might be more advantageous to rotate the very knowledgeable professors between universities--funded by NSF. Our problem may be technology transfer between universities.

3. The accreditation process might be a mechanism for putting pressures on state governments to support a minimum level in microelectronics.

4. There is some evidence that the Professional Licensing exams are encouraging the utilization of new technology.

5. There is evidence that business recognized its responsibility more than ever to support education. (Business Roundtable Report attached, April 1981) Let us develop a creative plan to utilize this feeling of responsibility.

6. The proposition could be advanced that an engineer in any discipline graduated today without some training in some of the tools in CAD is an incompletely trained engineer, i.e. without some familiarity with microprocessor systems, and interactive graphics.

Roundtable Report
No. 81-3*
April 1981

BUSINESS SHOULD INCREASE AND SUSTAIN PHILANTHROPY, GIVE IT GREATER IMPORTANCE

In a position paper on corporate philanthropy, The Business Roundtable has urged the business community to increase contributions for educational, health, welfare, and cultural activities and give philanthropic efforts greater importance.

The position paper was developed by a task force headed by Thomas A. Murphy, immediate past Chairman of the Roundtable and of General Motors Corporation. Both the National Association of Manufacturers and the United States Chamber of Commerce endorsed the Roundtable position.

While calling for increased giving by business, the Roundtable believes that decisions about the size and direction of contributions must be left to individual corporations since circumstances and needs differ from one company and one community to another.

Key points in the Roundtable statement include these:

--Philanthropy should be recognized as both good business and an obligation, so that businesses will be considered responsible corporate citizens of the nation and of the communities in which they operate.

--Business cannot survive and prosper unless the society in which it operates is healthy and vibrant. It is, therefore, in business' own self-interest, on behalf of its stockholders, employees, and various other constituencies, to engage in philanthropic activities which serve to strengthen the fabric of society.

--While the ability to make contributions tends to rise and fall with the course of the economy--and thus runs counter to times of greater or lesser social need--businesses should make an effort to maintain support of their most important philanthropic programs even during economic downturns.

--The principle alternative to private philanthropy is government funding. The sources of government funds, it must be emphasized, are tax-paying individuals and business enterprise. It is in everyone's

*Copyright 1981, The Business Roundtable. Reprinted by permission.

self-interest to support society through private social investments rather than through the complex and costly redistribution of tax dollars by government.

--If the business community is serious in seeking to stem overdependence on government, and still allow the not-for-profit sector to make the same contributions to society that is has in the past, business must itself increase its level of commitment.

--Businesses should establish appropriate programs to handle philanthropy in a businesslike way. Each firm has the responsibility to manage its philanthropic activities according to the same standards used to manage other parts of its business-- standards that apply not only to the quality of the individual managing the function, but that also involve the top corporate officers.

--Corporations frequently assist not-for-profit institutions through loans of employees, donations or loans of equipment and space, volunteer programs, and direct dollar investment in economic development efforts. These activities should continue and be increased.

--All companies should make public, in a manner they deem fitting, information on their philanthropic programs. Public dissemination of this information would serve as a means of increasing public awareness of the involvement of business in improving the quality of American life.

"LIMITS OF CORPORATE POWER" CATALOGS MULTIPLE RESTRAINTS PLACED ON BUSINESS

A penetrating examination of the existing restraints on the exercise of corporations' economic and social power is presented in a new book titled "THE LIMITS OF CORPORATE POWER" (New York: Macmillan Publishing Company, 265 pp., $19.95). The book counters the frequently the frequently-expressed view that corporate power is not subject to significant checks and balances.

The authors are Ira M. Millstein, an attorney who is a leading authority on antitrust law and trade regulations, and Salem M. Katsh, partner of Mr. Millstein in the New York law firm of Weil, Gotshal and Manges.

In five chapters devoted to as many classes of restraints on corporate power, the book analyzes:

--The limiting effects of corporate chartering laws;
--Limitations imposed by the law of supply and demand;
--Deterrents, as well as incentives, stemming from the tax system;
--Controls exercised by Federal and state regulatory statutes;
--The impact of social forces on corporate conduct.

The author believes it is vital that government "no longer play a hostile, antagonistic role vis-a-vis business, but instead recognize a mutuality of interest so that business and government can work together to achieve important economic and social goals."

In their view, this calls for recognition in government policy of the economic realities of today's world. While endorsing certainty of punishment for price-fixing and other competitve restraints, the authors suggest the ability of some sectors of the American economy to compete in world markets. Antitrust laws, they note, were created at a time when the U.S. was a self-contained market, without any real concern about the impact of foreign enterprises. They say there is little a corporation can do today, especially if it is a significant factor in a given market, "that is not subject to antitrust impressive array of overlapping enforcement programs and sanctions."

References for Microelectronics Talk
Engineering Education Conference, July 28, 1981
L. T. Rader

1. "Microelectronics," Scientific American September 1977 (whole issue).

2. "The Second Computer Revolution," (written November 1975,) Fortune 50th Anniversary issue, 2/11/80, page 231.

3. "VLSI - Implications for Science and Technology," by J. S. Mayo, Executive Vice President at Bell Labs. Presented at AAAS on January 4, 1980, San Francisco Annual Meeting.

4. "Issues in Engineering Education - A Framework for Analysis," National Academy of Engineering, Washington D.C., April 1980, 55 pages.

5. "VSLI goes to School," by Mason, Contributing Editor IEEE Spectrum November 1980.

6. "Analytical Summary and Perspectives on the Impact of Microelectronics on Productivity and Employment," by John W. Kendrick, George Washington University. Presented at special session of the OECD, Paris, November 27-29, 1979.

7. "Use of Technology for Teaching Said Lagging," The Chronicle of Higher Education October 22, 1979.

8. "The Microchip Revolution - Piecing Together a New Society," Business Week November 10, 1980, page 97.

9. "The Coming Impact of Microelectronics - Joining Hands Against Japan," Business Week November 10, 1980, p. 96 & 108.

10. "Technologies for the 80's," Business Week July 6, 1980, page 48.

11. "Those Worrisome Technologies," Fortune May 22, 1978 (a good chart on where the U.S. leads in innovating)

12. "The Ultimate Frontier," Lewis Branscomb, Chief Scientist and Vice President of IBM. Condensed version of speech made at a General Electric symposium in 1980.

13. "Impacts of Microcomputer Technology on the Management of Product Development," by Kalinowski of Corning Glass Company, 1980. IEEE Engineering Management Conference Record, catalogue number 80CH1603-0, page 52.

14. "College CAD-CAM Consortium," Virginia Engineering Review Volume 7, number 1, Winter 1981.

15. "Professors' Conference News Letter," General Electric Company, May 1981. Discussion of new Microelectronics Center in North Carolina.

TEACHING OF ENGINEERING COURSES WITH STUDENT FACULTY RATIOS FAR IN EXCESS OF PAST EXPERIENCES

John C. Lindenlaub
Professor of Electrical Engineering
Purdue University
School of Electrical Engineering
West Lafayette, Indiana 47907

Summary

This discussion is centered around the problems which have arisen as a result of teaching engineering courses with student faculty ratios far in excess of past experiences. A representative list of problems as perceived and expressed by engineering faculty members is presented. It is pointed out that engineering undergraduates tend to see many of the same problems with large classes. Typically, an engineering professor is not faced with one or two of these problems but with all of them simultaneously. As a result the time and effort required to meet day to day teaching commitments is beginning to take its toll in terms of long-range faculty development. The problems brought upon us by large classes, coupled with associated problems relating to faculty shortages, competition for faculty members by industry, and a decline in graduate enrollments - particularly at the Ph.D. level - has placed engineering education in a crisis situation.

In an effort to overcome these problems it is recommended that we continue to lobby for more resources and control over engineering enrollments, that we take advantage of the opportunities offered by peer teaching, that we seriously rethink the role of the undergraduate engineering educator, and that we learn to use educational technology more effectively in engineering education.

WE WANT OUR STUDENTS TO:

    LEARN BASIC FACTS AND PRINCIPLES

    DEMONSTRATE COMPREHENSION

    APPLY THESE IDEAS IN A ROUTINE MANNER

    ANALYZE - KNOW WHEN TO APPLY WHAT

    SYNTHESIZE - OPEN ENDED PROBLEMS, DESIGN

    EVALUATE - CRITICAL ENGINEERING JUDGEMENTS

FIGURE 1
TYPICAL EDUCATIONAL GOALS IN ENGINEERING COURSES

IN PERSONAL ATMOSPHERE
- GETTING TO KNOW STUDENTS
- HANDLING QUESTIONS
- DEALING WITH STUDENTS INDIVIDUALLY

ABILITY TO EVALUATE FAIRLY
- HOMEWORK
- TESTS
- MAINTAINING STANDARDS
- MIXED BACKGROUND OF STUDENTS

STUDENT DISCIPLINE
- ATTENDANCE
- CHEATING

FACILITIES AND INSTRUCTIONAL AIDS

LOGISTICS
- PASSING OUT AND COLLECTION OF MATERIALS
- TRANSPORTATION OF MATERIALS
- MECHANICS OF TESTING
- LATE HOMEWORK, MAKEUP EXAMS
- STAFF COORDINATION
- DIFFICULT TO "SHOW AND TELL"
- MAINTAINING STUDENT RECORDS

NO TIME TO INNOVATE

AVAILABILITY OF PROFESSOR OUTSIDE OF CLASS

LIMITED LAB SPACE AND EQUIPMENT

FIGURE 2
TYPICAL LIST OF PROBLEMS ASSOCIATED WITH LARGE ENROLLMENT
CLASSES AS PERCEIVED BY ENGINEERING FACULTY

PROVIDE MORE RESOURCES

LIMIT ENROLLMENT

EFFECTIVELY USE PEER TEACHING

INTERDISCIPLINARY TEAM TEACHING

MAKE EFFECTIVE USE OF EDUCATIONAL TECHNOLOGY

FIGURE 3
SOME APPROACHES TO DEALING WITH ENGINEERING EDUCATION'S
LARGE ENROLLMENT PROBLEMS

"WHY ARE ENGINEERING STUDENTS NOT BEING TAUGHT TO WRITE
and SPEAK WELL, and WHAT ARE WE GOING TO DO ABOUT IT?"

Ralph E. Jenkins
Associate Professor of English
Temple University
Philadelphia, Pennsylvania 19122

Summary

In our present educational system, engineering students are increasingly likely both to enter and to leave with deficient communication skills. Consideration of human language as an information storage and transfer system shows us that the problem has several sources: the inherent difficulty of changing ingrained language habits, the lack of variety in communication opportunities for engineering students, inappropriate training of teachers, and inadequate cooperation among schools and universities. Existing remedial and composition programs can help with certain problems; for others, professional writing courses and experiences that provide a wide variety of situations, purposes, and audiences will be necessary. There are not enough qualified teachers for such courses at present because typical English teachers do not have the appropriate attitudes, information, and experience. Current research and practice suggest that we can provide adequate teaching if we pursue research into the causes of specific writing problems, provide students with varied opportunities to communicate with varied audiences, provide communication, set up cooperative workshops and personnel exchanges between industry and universities, and encourage coordination of high school and college teaching.

Because of the limitations of time, I am going to make some broad generalizations about engineering students, English teachers, language, and the problems of teaching communications skills. I therefore want to sketch out my experience so that you can judge how far to trust these generalizations. In college, I started out in engineering and worked for seven years in a steel fabrication company designing storage tanks and structural steel. I then got a Ph.D. in English literature. For the past five years, I have taught technical writing; I also organized and managed Temple University's freshman composition program; I now administer our remedial writing program and direct our center for research into learning problems of students in remedial mathematics and communication skills. A large part of my job is training people with backgrounds in literature to become teachers of writing.

Our collective goal, as I see it, ought to be to ensure that engineering graduates are able to write and speak confidently and effectively about the problems, methods, consequences, results, and value of their work to a wide variety of audiences in a wide variety of situations. The fact is that many, perhaps most, engineering graduates do not have these abilities and are not likely to develop them in our present system of education, and the situation may very well get worse. Every year more and more students enter college with deficient communication skills; the job market attracts more and more of them to technical fields; and more and more foreign students with minimal English skills come to us for technical education. Let us look at the problems with our present system and see what can be done about them.

First, let us quickly review some characteristics of human language considered as a communication system. By <u>communication</u> I mean here the intentional re-creation in some mind or minds of an experience that originally occured in some other mind. By <u>technical communication</u> I mean the flow of such communication within an organization that enables the organization to produce the results for which it was designed. Language is an information storage and transfer system that enables us to communicate solutions to problems of individual and species existence across space and time. As a species, we have been speaking for hundreds of thousands of years, and our brains are genetically programmed for speech. Speech, however eacy to learn, is difficult to transfer across large distances, and very difficult to store in long-term memory, especially when its content is precise technical information. As Eric Havelock points our in <u>Origins of Western Literacy</u>, the recording of original thought, and consequently modern science and technology, were not possible until efficient systems of inscribing words and numbers were invented -- specifically, the Greek alphabetic script and the Indo-Arabic mathemitical notation. As a species, we have been writing in any script for less than 5000 years, and our brains are not genetically programmed for writing as for speech. Writing calls for a complex and somewhat unnatural association of eye, ear, brain, and hand that must be developed early in life if it is to be automatic; few of us encounter cognitive and emotional problems very different from those involved in learning a body of information or set of concepts.

Writing and speaking, considered as communication, differ from cognitive acts such as working out a mathematics problem in that communication requires the participation of at least one other person. In a given situation, an act of human discourse requires a speaker or writer, a listener or reader, the material to which the discourse refers, and the form of the discourse itself - the written or spoken message, which is a set of symbols chosen from the possibilities afforded by the language and script being used. The form of the discourse is thus determined by five things: the range of possibilities inherent in the language and script, the situation that calls for the discourse, the material referred to, the writer or speaker's purpose, and the needs and purposes of the audience. The act of communication is not complete when the discourse has been formed; it is complete only when the message has been received and its content and function understood and acted upon. This means that when communication is impaired, there may be several possible causes. The problem may be that the language or script writer, or in the reader, or in both, or it may be that the language or script chosen for the message is inadequate to carry the intended meaning. Locating and correcting faults in human communication is difficult because of the complexity of the system and because of the emotions of writer and reader. Becuase language to a large extent determines our sense of self and reality, we often feel threatened when confronted with a language deficiency; we may refuse to acknowledge it or try to shift it onto someone else. The poor writer blames his reader, and vice versa.

These considerations mean that an effective system for teaching communication skills must be capable of cutting through unconsciousness, resistance, and pretense about communication to locate and correct any of these possible problems. Student writers, for example, may lack fluency; they may show little awareness of situation, audience, or purpose; they may have difficulty finding, organizing, or explaining material; they may construct garbled sentences, or use non-standard grammar, or be unaware of the conventions of alphabetic script; they may have neurological or motor defects that impede the production of speech or script. Or the source of the problem may be the reader, who may be unaware of his deficiencies or refuse to acknowledge them. If the reader is also a teacher, the problem is doubly complex, and is likely to be analyzed and solved only by outside intervention. For this reason improving the teaching of communication skills is only partly a result of advances in theory or technique: it is also dependent upon transformations in the experience of students and teachers. Both teachers and students must be willing to overcome their own limitations to find or create a common ground, a common language, a shared purpose, before any theory or technique can be effective.

Advances in theory and technique have been made, of course. To be brief, composition teachers have found that students' skills develop more quickly if they communicate for real purposes to real audiences, if the teacher stresses fluency before form and form before correctness, and if the teacher participates in the class activity rather than stand aloof as instructor and judge. At Temple, we are finding that common errors in writing may correlate more with the brain mechanisms involved in text production that with the standard categories of grammar handbooks, and ar drawing analogies between computers and human language and script that may make error correction more efficient. However, this is not the place to review current knowledge, and despite these advances most composition teachers at present know less about their subject than the average television repairman knows about his. A TV repairman understands the structure and function of the TV set, and its relationship to the broadcasting system, and can thus form and test hypotheses about the causes of problems of transmission and reception until the problems are eliminated. Comparable knowledge for composition teachers would be general theory of the social, technical, and psychological origins of writing, an understanding of its operating principles considered as a information system, and understanding of the neurological mechanisms and cognitive processes involved in the production of speech and text, hypotheses about the causes of specific writing problems, and specific methods for eliminating them. In the absence of such knowledge, composition teachers routinely instruct their students to imitate desirable forms of writing, point out their errors, congratulate themselves if the students improve, and blame the students if they don't, labelling them as innately deficient. This move is particularly tempting if a whole class of students, such as engineers, already has a reputation for writing badly.

Of course, it may be true that because of neurological problems or lack of early training, some students may never be able to use language well enough for some professions; we are not all equally adept at using symbol systems. However, this is not likely to be the case with graduating engineers, who must have demonstrated competence in two or three other symbol systems: mathematics, computer languages, and engineering graphics. For these students, lack of communication skill is likely to be the result of inadequate teaching. Adequate teaching must go beyond correction of errors in script and grammar, and beyond imitation of desirable forms, to give the students a grasp of the structure and operating principles of script and discourse, practice in analyzing the demands of different situations and the needs of different levels of readers, and practical experience in manipulating the variables that determine the form of discourse (that is, situation, audience, purpose, and material). The students should be encouraged and required to write and speak about technical material to audiences of different technical levels, in a variety of situations, for a variety of purposes, and should be given opportunities to compare the intended effect of their communication to the actual effect on genuine audiences, and to correct where necessary.

Compare this ideal sketch to the actual experience of engineering students. Temple's program may be typical. If the students have severe problems of fluency, invention, organization, syntax, grammar, or script, they go into remedial writing or speech until these problems are substantially corrected. They then go into freshman composition; if the course is well designed and the teacher well trained, the students will develop the ability to write effective essays on subjects of general interest. However, no matter how good they are of their kind, these courses will not provide the variety of audience, purpose, and situation needed to develop professional skills, and will not address the special problems of technical communication - for example, the need to describe microscopic or astronomical reality in the language of ordinary human experience for lay audiences. Suppose, then, that we rely on the students' experience in engineering courses to provide these professional skills. We discover that most of them spend their study time on the technical aspects of engineering, read only textbooks, talk only to fellow students who take the same courses, and write only to professors who know more about the subject than they do. In this closed circle there is little or no variety of situation, audience, or purpose; the students can communicate successfully in the kind of shorthand that is possible when speaker and listener have identical backgrounds and interests. The only determinant of form that changes for them is material so it is not suprising that outsiders complain that engineers care only about technical material and are innattentive to audience and situation. After all, their experience has not required them to respond genuinely to the different needs of different audiences, and they may have written for only one purpose - to be evaluated by their professors.

Suppose, then, that we decide to require an additional course in technical writing, as we do at Temple. Who teaches the course? Keeping in mind that in genuine communication, the reader or listener is as important as the writer or speaker, let us see who the student's reader listener, and teacher is likely to be.

Typical university English teachers were trained in leterature and critical theory, with perhaps some linguistics. They want to teach literature courses in their field, preferably graduate courses, and regard the teaching of writing as an unavoidable evil. Their background has led them to confuse literature with the whole of writing, so that they regard all other forms of writing as inferior to poetry and fiction. They are well trained in the analysis and evaluation of text, but have no formal training in the production of text. They regard all texts primarily as static forms to be appreciated aesthetically if possible, and disregarded otherwise, and are unaccustomed to thinking of purposeful messages designed to enable an active reader to produce a specified result in the physical world. They are ignorant of mathematics and graphics, are both fascinated by and frightened of computers, and regard any technology above the level of electric can opener with suspicion and distrust. They have no business or technical experience, and thus no knowledge of the function of writing in large organizations. On the one hand, they secretly regard engineers and businessmen as barbarians, and themselves as defenders of the ramparts of western civilization; on the other, they are fascinated by what they perceive as the power of industry. On the positive side, they are usually intelligent, dedicated to scholarship and teaching, sensitive to nuances of language and behavior, and often educable if approached in the right way.

This description may sound exaggerated, but I assure you it is a considered conclusion from years of observation. There are of course some exceptions, especially among the younger faculty who cannot get permanent jobs, but not enough to staff the growing number of technical and professional writing courses.

Let us now sum up what we know about the sources of the problem of teaching effective technical communication.

1. Students have barriers to writing well and speaking well. Writing is inherently difficult, and speech habits are deeply ingrained. Some students are not native speakers of English; some speak non-standard dialects; many had little or no speech or composition training in school; most resist criticism of their language use, especially when they believe the critic to be incompetent to understand their material. In addition, their educational experience does not provide the frequent and varied opportunities for communication and response that might encourage natural development of skills.

2. Teachers have barriers to teaching well. Their training has not prepared them to teach functional writing. They are insecure about business and technology because of their ignorance, and defensive about the value of literature and liberal arts. They do not know the causes of writing problems, and are likely to try to teach writing by imitation of forms, error correction, and blaming.

3. Our educational system is not well organized to attack the problem. There is insufficient cooperation between high school and colleges. Particularly in urban schools, few teachers can

effectively teach speaking and writing because of lack of training and lack of institutional, family, and social support. Few universities have graduate programs in the teaching of composition or pursue research into communication problems, or give permanent jobs to composition specialists.

What can we do about this situation? Here are four recommendations.

1. Universities, industry, and government can fund <u>research</u> into the causes of actual communication problems and patterns of skill development. Of particular interest are the problems faced by speakers of other languages and non-standard dialects who must learn Standard English at the level of accuracy needed for clear technical communication, and the writing problems faced by students from the other script systems, such as the Middle Eastern syllabic and the Chinese logographic systems. This research will allow us to diagnose problems and place students more accurately; the results will improve the efficiency of existing remedial programs, especially those for entering minority and foreign students and can be used in the design of efficient support systems, such as computer-assisted tutorial programs.

2. Universities can provide engineering students with <u>opportunities to practice communication skills</u> in a variety of situations. Some of this can be done by requiring technical courses, but a single course is not enough. Composition and speech faculties can set up workshops for assisting engineering faculty in teaching and evaluating communication skills, so that engineering faculty can require and evaluate regular oral and written presentations in their own courses. Engineering, Speech, English, and Business faculties can cooperate in the design of communications programs to provide students with a variety of audiences and responses from non-engineers.

3. Industry and English departments can provide English teachers with the <u>information and experience</u> that will enable them to teach functional writing. Teachers need information about the forms and functions of technical communication comparable to the information they now have about the forms of literature. Engineering corporations can give researchers from English departments the opportunity to apply their scholarly and critical training to the study of the internal and external communications of the corporation; access to some files and interviews with engineers and technical writers will give researchers valuable knowledge of their subject. Teachers also need experience in the activity they teach. Engineering corporations and English departments can cooperate in setting up workshops for teachers and exchanges of personnel. Corporations can use graduate students or technical writing teachers as editorial interns; English departments can use engineers or technical writers as adjunct faculty.

4. School districts, colleges, and universities can <u>coordinate</u>

their efforts. Through existing networks such as the National Writing Project and its local sites, universities, corporations, school districts, and funding agencies can make the results of this research and experience available to teachers in school systems to improve the teaching of communication skills in high schools.

All these recommendations, I think, are possible and realistic, but they take an unusual effort of coordination. English departments will have to give up some of their traditional monastic defensiveness; engineering corporations will have to recognize the contribution that literature teachers can make to their work; universities will have to give up some of their pretended isolation from the school districts they serve; and we will all have to come up with some money. If we do, we can get the result we want. If we don't, we may find that the next generation of engineers is rebuilding the Tower of Babel.

KEY ASPECTS OF COMPARISON BETWEEN ENGINEERING EDUCATION
IN THE USA AND ENGLAND, JAPAN, GERMANY, AND RUSSIA

by

J.S. Przemieniecki
Dean, School of Engineering
Air Force Institute of Technology

The United States has benefited in the past by adopting educational models that were being used in countries in Europe. Engineering curricula in this country initially were patterned after French models, particularly that of the Ecole Polytechnique, to the extent that most of the first engineering textbooks were either French or translations from the French. The original colleges and universities in the United States which introduced engineering curricula reflected naturally a British influence with the traditions of Cambridge and Oxford universities. Thus the early engineering education was a mixture of the French and British systems. Subsequently, as the engineering schools evolved further, the impetus for research and graduate education came from Germany.

We should not, however, infer from this that the engineering education developed in the United States totally as a result of borrowed elements. These early ideas were adopted into American culture with its desire to promote "the applications of science to the common purposes of life," which gave a unique character to engineering education. This process led to the development of many creative ideas in engineering education such as the establishment of the land-grant college, the teaching laboratory, cooperative education and, in more recent years, the engineering management programs. Many of these concepts, in turn, have influenced the development and evolution of engineering education in other countries throughout the world.

In the past few decades, however, with the tremendous advances America has achieved in science, technology, and higher education, we have become insular as a nation with respect to outside ideas. These advances, including the dominance of the Nobel Prizes, have prompted other countries to emulate US approaches and adopt US educational models. These same achievements, however, may have caused Americans to abandon examining seriously the educational systems of other countries and learning from their experiences.

Technical and educational leadership cannot be maintained without continued creative thought and appropriate changes. This can be fostered by a study and better understanding of the efforts of others, but it is not suggested here taht we should mirror image their efforts. Rather, we should study what other highly industrialized countries are doing since their concepts and ideas may assist engineering educators in the United States to think more imaginatively about problems facing the national economic well-being in general or the engineering education in particular.

In this paper a general overview of the educational systems as they relate to the engineering education, in the United Kingdom, Germany, Soviet Union, and Japan is presented. In addition, two specific recommendations are made for consideration by the Engineering Foundation Conference on the "Engineering Education: Goals and Aims for the Eighties."

United Kingdom (Figure 1)

Following a 4-year program in the Junior School, children in the United Kingdom (UK) enter the Secondary School for a period of 5-7 years. At the end of the fifth year in the secondary school they are expected to take up to 10 General Certificates of Education at the Ordinary Level (O-Level) while in the fifth and sixth year they can take up to four General Certificates of Education at the Advanced Level (A-Level). At this point they may enter a technical college and qualify as an engineering technician. Several different levels of certification are possible for students who choose this path. Following a period of practical experience they can take Council of Engineering Institutions examination to qualify as Chartered Engineers (C. Eng.), title akin to that of a "Professional Engineer" (PE) in the United States. On the other hand, students who pass at least two A-Level examinations have a mandatory right to a local educational grant to support them and pay their fees during at least three years of further education. Thus the UK educational system operates under the policy that the higher education should be available to all those who are qualified and willing to undertake it.

The other path to engineering education can be followed either at a university or a polytechnic institute (polytechnics). Each university has its own Royal Charter and they have more autonomy than the polutechnic institutes. The University Grants Committee, which provides funding for the universities in England, does however exert very strong influence on the range and type of programs. Polytechnics are controlled by the Council for National Academic Awards and they are funded by local educational authorities. Because of their charter, polytechnics are much more innovative in the development of new programs or in curricular changes compared with the universities which follow a more conservative path.

Bachelor's degree is obtained after three years except in the so called "sandwich program" (equivalent to cooperative programs in the US) where the program is four years in duration. More recent trends in the engineering curricula have been in the direction of increasing the content of humanities, social sciences, computer technology and computation, and project design. Students are allowed to take more electives than it was possible under the previous modular structure. In general, the curriculum content complies very closely to the criteria of the Accreditation Board for Engineering and Technology in the United States.

The Master's degree can be obtained after one year of study which must include a project or dissertation. There are two types of doctorates granted in the United Kingdom: the Doctor of Philosophy (Ph.D.) and the Doctor of Science (D.Sc.). The Ph.D. may be obtained through one or more years of study and research beyond the Master's degree A dissertation which makes a significant contribution to the existing knowledge is required. The Ph.D. degrees are granted by both the universities and polytechnics.

The D.Sc. degree is a higher doctorate and is granted only by the university on the basis of a review of all candidate's contributions, including books, papers and reports. In some cases, before the degree is granted, the candidate may be required to complete additional work.

Federal Republic of Germany (Figure 2)

Compulsory education begins at 6 years of age in a Grundschule for four years. After Grundschule a choice of three types of secondary schools is available: Hauptschule, Realschule, and Gymnasium. The major differences between schools are summarized below:

The Hauptschule - The program length is 5-6 years and normally this school will not allow progress to higher education. Students from this school go to vocational training and become certified as technicians.

The Realschule - This school is academically less demanding than the Gymnasium. After the Realschule (6 years) students continue in the Fachoberschule (2 years) followed by the Fachhochschule (2 years) where they graduate with the title of the Ingenieur Graduiert. This qualification is equivilent to the Bachelor of Engineering Technology used in the United States.

The Gymnasium - This school is the most demanding of the three. Students receive a diploma (Abitur) and they have right to enter a university.

Parents have a choice of one of the above three secondary schools where they want to place their children, but after two years the Education Ministry can transfer any student to a type of school it deems more appropriate to the child's ability. Attempts are underway in the Federal Republic of Germany to delay the secondary school selection and give students a wider option regarding their future careers. Also a new type of school, the Gesamtschule, is being created with the hope that it will replace all three secondary schools presently used.

Following graduation from the Gymnasium there are four separate paths to qualify as an engineer in the Federal Republic of Germany. The first path is through a four year program at the Gesamthochschule, where depending on the type of program students graduate with the title of Deplom Ingenieur or Ingenieur Graduiert. The second path is to go through the track including the Fachoberschule and Fachhochschule. The other two are either through a university (Universität) or Technische Hochschule. In both cases they graduate with the title of Diplom Ingenieur. After additional two to three years of full-time study, culminating in a thesis and an oral examination, the student may be awarded the degree of Doktor Ingenieur.

Students after obtaining a doctorate often continue as professional assistants doing research for several additional years to prepare a second thesis of a more fundamental nature which must be defended before the entire faculty. When this is done, the person is deemed prepared to teach and is given the title Doktor Habil.

Union of Soviet Socialist Republics (Figure 3)

Prior to World War I there were few schools in Russia and almost 75% of the people were illiterate. Higher education was limited to a privileged class only. Today ten years of schooling is virtually the norm. The general school system consists of four-year elementary schools, eight-year schools and ten-year schools. Students who complete the four-year school are transferred to either the eight- or ten-year schools. Completion of the eight-year school leads to vocational and specialized training at vocational schools or specialized secondary schools. The ten-year school prepares students to enter academic institutions of higher education.

There are two types of academic institutions: universities and institutes, the latter are referred to as either polytechnic institutes or specialized institutes (e.g. Aviation Institute in Moscow). Admission standards to highly desirable institutions (e.g. in and around Moscow) are very high. Of the two types, the universities are relatively small in number representing only 8% of the total number of institutions and 13% of the total enrollment.

The engineering programs at these institutions extend over 5 to 5½ years. They are approximately equivalent to the Master's degree level in the United States. The curriculum consists of:

2 years of general scientific training
2 years of theoretical fundamentals in the chosen speciality
1 year of specialization in the major field of study
½ year of project design which must be defended before a State Examination Commission. Project is also reviewed by a representative from industry.

Mathematics and social sciences are included throughout the whole period of study. Students are required to take 400-800 classroom hours of mathematics and a foreign language course is mandatory.

Engineering curricula are more specialized than in the United States. There is , for instance, distinction made among electrical engineers, electromechanical engineers, and electrophysics engineers. In recent years, however, there has been a tendancy to broaden the specialties and make them more general.

The first advanced degree is the Candidate of Science (Kandidat Nauk) or Candidate of Engineering Science. This degree requires three years of study, dissertation, examination in a foreign, a social science, or humanities subject, and his area of specialization. The results of the dissertation must first be published (usually 2-3 papers). For the defense two official opponents of the dissertation are appointed to critically review the thesis and contest its claims. The degree must be confirmed by the Supreme Certifying Commission (VAK) before the university or institute can award it to the student.

The second advanced degree is the Doctor of Engineering Sciences (Doktor Nauk). The basic requirement for this degree is a second dissertation which must contain significant scientific or engineering results such as initiating a new direction for research and development. The content of the dissertation corresponds roughly to a monograph largely based on original results. The Doctor of Engineering Sciences degree is awarded by the Supreme Certifying Commission.

Japan (Figure 4)

The present educational system in Japan follows very closely that of the United States. Six years in the elementary school, followed by 3 years in the Junior High School and then either 3 years of the General High School or the Vocational School. Admission of students to the university or institute of technology is by examination. There is a very keen competition to get into prestigious universities. The engineering curricula include:

$1\frac{1}{2}$ years of basic science, including mathematics
$2\frac{1}{2}$ years of specialization, including a project which is of an experimental or analytical nature, but rarely an engineering design.

There are very few electives and all students follow a rigid curriculum.

Recommendation

1. Studies of engineering curricula in the European countries and Japan should be undertaken with the view of obtaining a better basis for comparison of the US educational system and in other countries. Such studies could perhaps be conducted under the overall guidance of ABET or AAES. The findings of such studies would be useful in planning future directions of emphasis for the engineering education in this country.

2. The feasibility of establishing a higher doctorate in engineering somewhat similar to the Doctor of Sciences in England, Doktor Habil in Germany, or Doctor of Engineering Sciences (Doktor Nauk) in the Soviet

Union should be considered. This new degree would be an earned doctorate, not to be confused with the various honarary degrees presently awarded by American universities. A suggested set of criteria for this degree and the main advantages are listed in a separate appendix to this paper.

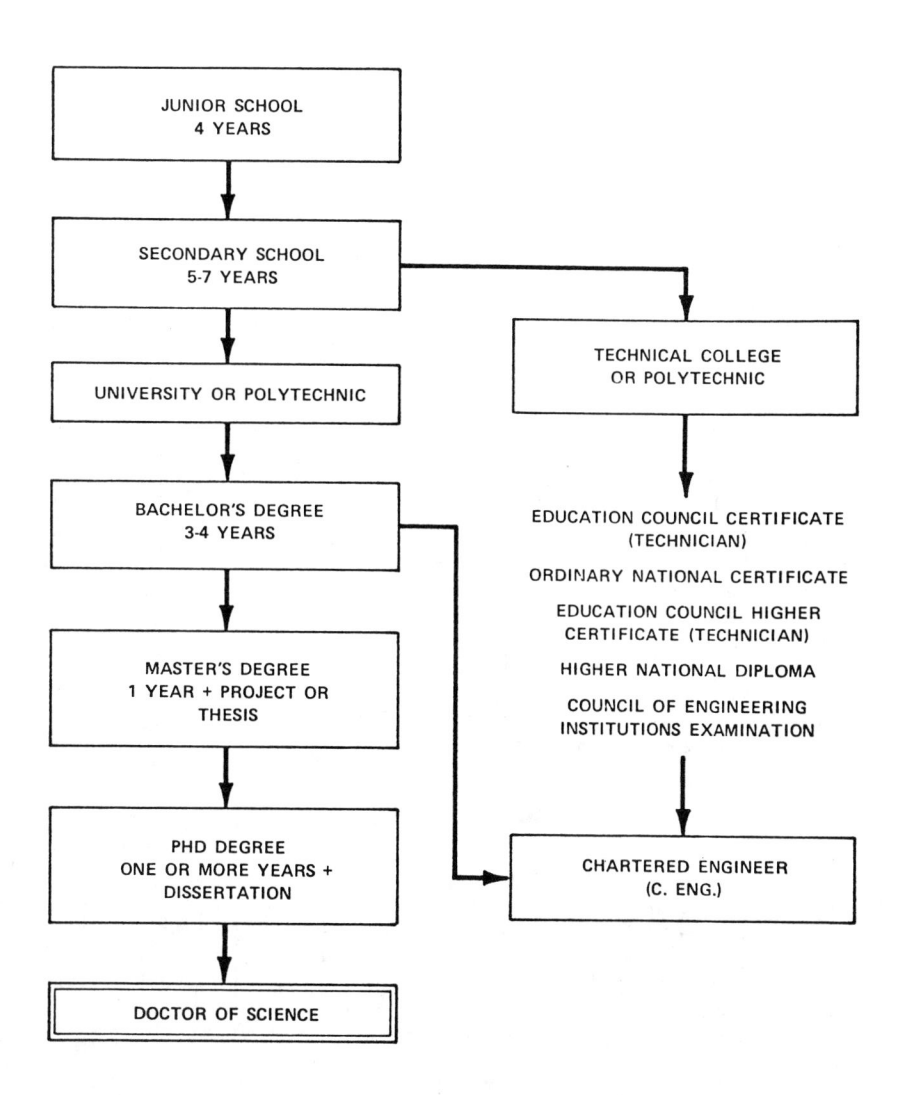

```
┌─────────────────────┐
│  JUNIOR SCHOOL      │
│    4 YEARS          │
└─────────┬───────────┘
          ▼
┌─────────────────────┐
│  SECONDARY SCHOOL   │───────────────┐
│    5-7 YEARS        │               │
└─────────┬───────────┘               ▼
          ▼                ┌─────────────────────┐
┌─────────────────────┐    │  TECHNICAL COLLEGE  │
│ UNIVERSITY OR       │    │   OR POLYTECHNIC    │
│ POLYTECHNIC         │    └──────────┬──────────┘
└─────────┬───────────┘               ▼
          ▼                EDUCATION COUNCIL CERTIFICATE
┌─────────────────────┐    (TECHNICIAN)
│  BACHELOR'S DEGREE  │───►
│     3-4 YEARS       │    ORDINARY NATIONAL CERTIFICATE
└─────────┬───────────┘
          ▼                EDUCATION COUNCIL HIGHER
┌─────────────────────┐    CERTIFICATE (TECHNICIAN)
│  MASTER'S DEGREE    │
│  1 YEAR + PROJECT   │    HIGHER NATIONAL DIPLOMA
│    OR THESIS        │
└─────────┬───────────┘    COUNCIL OF ENGINEERING
          ▼                INSTITUTIONS EXAMINATION
┌─────────────────────┐               │
│   PHD DEGREE        │               ▼
│ ONE OR MORE YEARS + │──► ┌─────────────────────┐
│   DISSERTATION      │    │ CHARTERED ENGINEER  │
└─────────┬───────────┘    │     (C. ENG.)       │
          ▼                └─────────────────────┘
┌─────────────────────┐
│ DOCTOR OF SCIENCE   │
└─────────────────────┘
```

Fig. 1 THE EDUCATIONAL SYSTEM OF THE UNITED KINGDOM (UK)

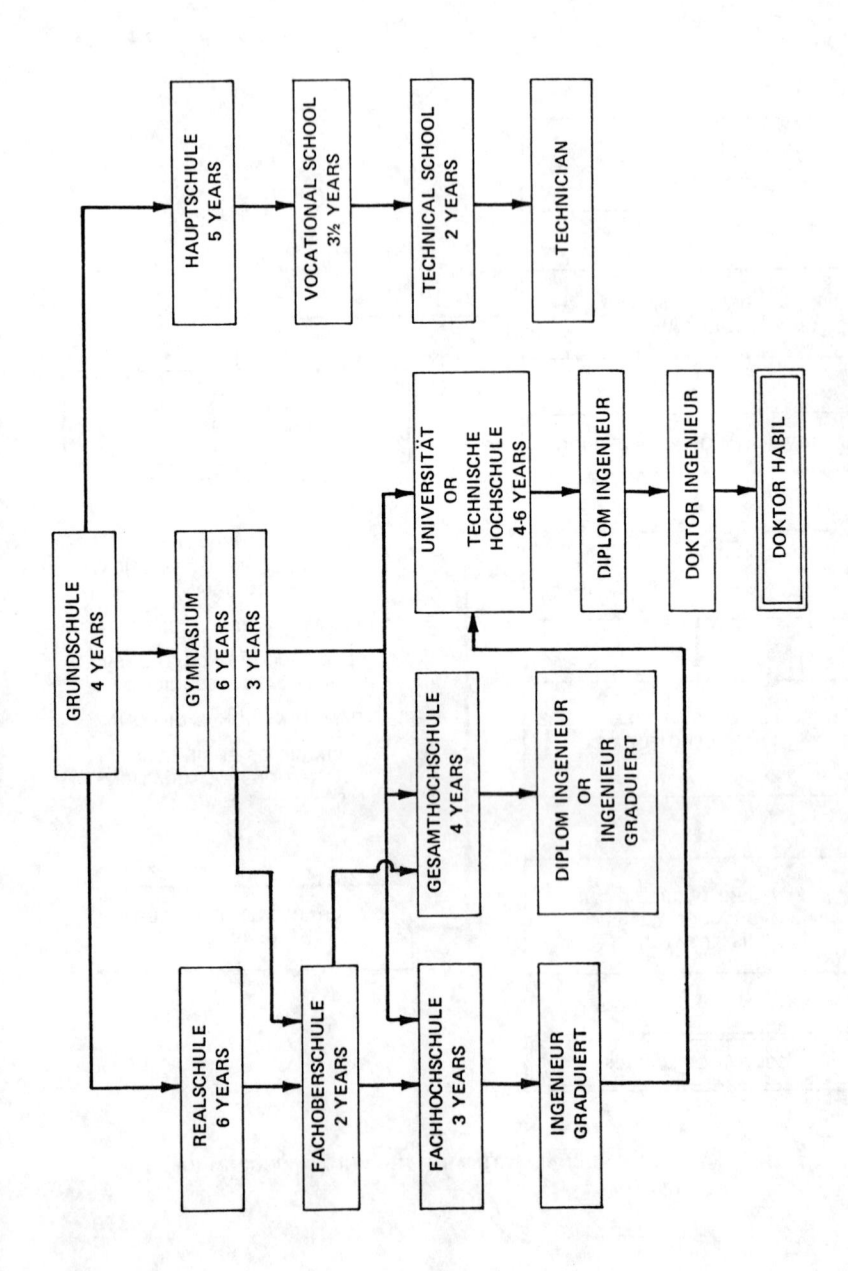

Fig. 2 THE EDUCATIONAL SYSTEM OF THE FEDERAL REPUBLIC OF GERMANY

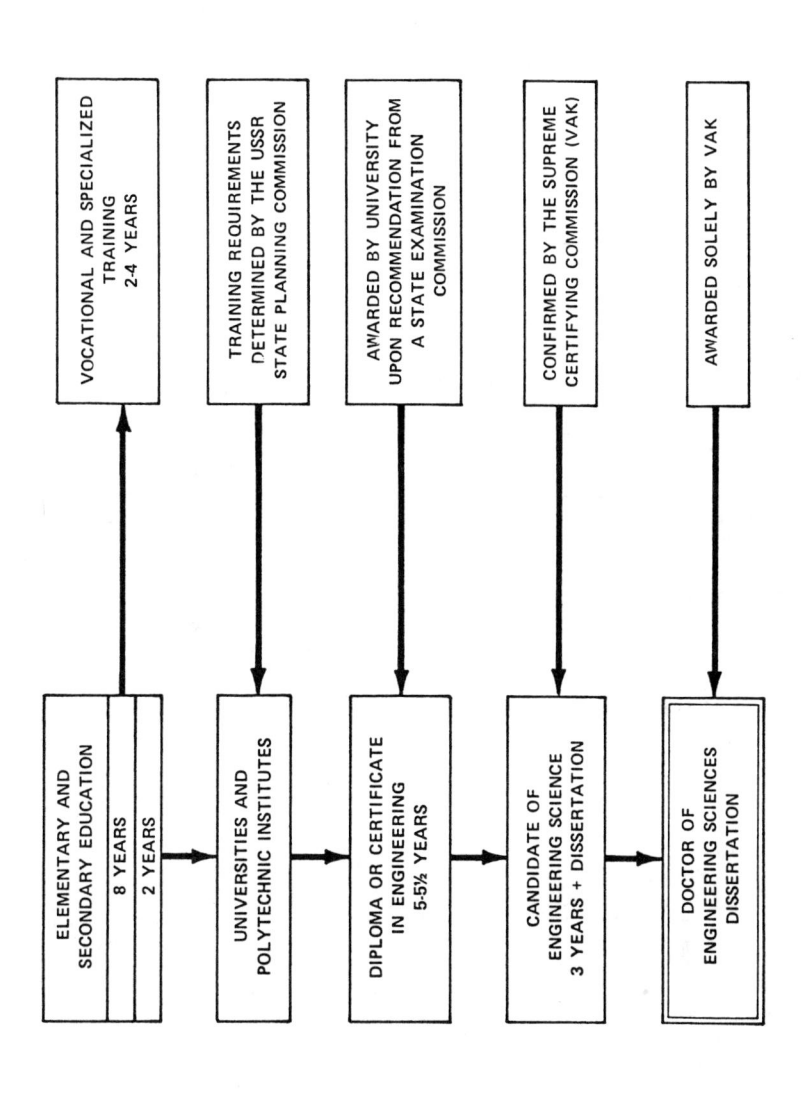

Fig. 3 THE EDUCATIONAL SYSTEM OF THE SOVIET UNION

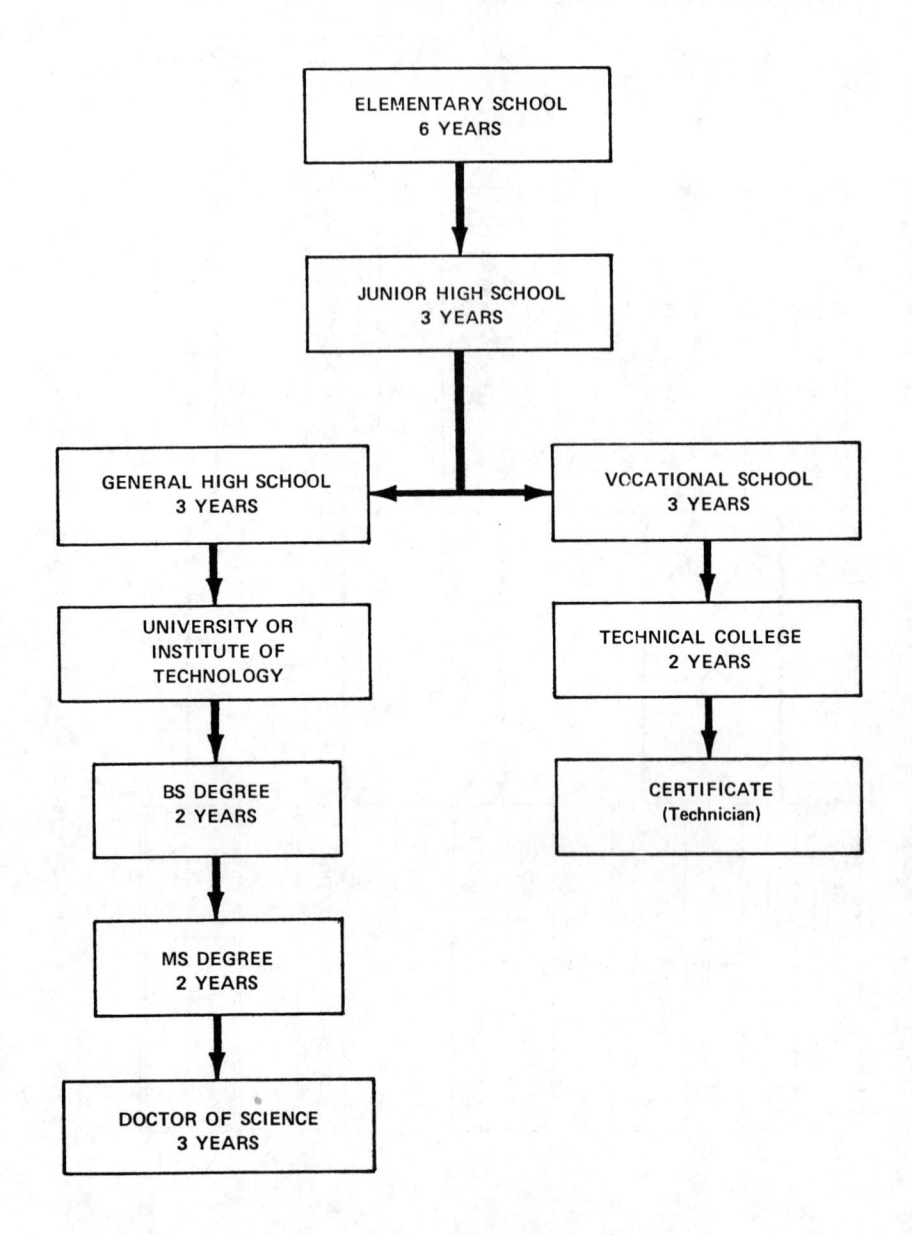

Fig. 4 THE EDUCATIONAL SYSTEM IN JAPAN

HIGHER DOCTORATES IN ENGINEERING

CRITERIA

* Higher doctorates would be awarded only to holders of graduate degrees (Typically Ph.D.'s).

* Potential candidates will apply to the university to be considered for higher degrees. If approved, the university will appoint a special Examining Board of Experts selected from the following:

  * Senior faculty members (within and outside university).

  * Holders of higher doctorates in the same field of specialization (in future).

  * Members of the National Academy of Engineering or National Academy of Sciences.

  * The Graduate Dean (ex officio).

* Candidates must demonstrate significant contributions to the existing knowledge through original research (e.g. published papers, reports, books, monographs, etc.).

* Candidates must achieve recognized standing among their peers.

* Candidates will normally be required to complete a _monograph_ on the subject agreed by the Examining Board.

ADVANTAGES

* Stimulate fundamental contributions to the engineering science and design at the post-doctoral level (published monographs).

* Provide additional academic recognition by the university to outstanding scholars in engineering fields.

* Provide excellent motivation for sabbatical work.

CRISIS IN AMERICAN ENGINEERING EDUCATION

by

Jerrier A. Haddad
IBM Vice President
Technical Personnel Development

CRISIS IN AMERICAN ENGINEERING EDUCATION

There is a June 28 New York Times Magazine cover story, "The Technology Race." Let me quote from that article, because it deals quite directly with the subject at hand.

"Floyd Humphrey (Head of Electrical Engineering Department at Carnegie Mellon University) had scarcely unpacked...before he began encountering the problems that so typify the nation's receding position in world technology." What were these problems? Seniors weren't attending class because they were out interviewing for jobs. The average senior had 15 interviews. Only seven out of 60 in the master's program were staying to work on a doctor's degree, and his department needs at least 20 doctoral candidates to survive. Faculty members' research projects will wither without graduate students. The exodus of faculty, already started, will increase. Therefore, research grants will decrease, making it harder to attract new faculty and students."

The second quotation comes from a boxed summary called "The Technology Gap":

"During the last decade the number of doctorates declined by one-third. And one-third of these awarded to "temporary resident's" or foreign students...

...In 1979, for the first time, American productivity actually declined by almost on percentage point.

...In the last decade, while undergraduate engineering enrollment was growing by 47 percent, the number of doctorates was down one-third.

...In 1980, an estimated 10 percent to 15 percent of the nation's engineering faculty positions were unfilled.

...In effect, the farmer is eating his seed corn instead of planting to get a new crop."

There's no shortage of suggestions as to what's the fundamental underlying cause of the impending disaster. There's no shortage of identified villans on whom the crisis can be blamed. You can blame industry; you can blame government; you can blame politicians, educators, the President. The list can go on ad infinitum. Among the choices for the real problem, the depth problem, are these.

* The American public school education system is in bad trouble because they don't teach science; they don't teach mathematics, and yet, at the same time, I've got to observe that I hear you gentlemen say that you're in front of classes of superb students with none better trained.

* There's a decline in the real dollars available for R&D. Many presentations have been made about this decline in R&D spending as a fraction of GNP relative to our competitive nations of Japan and Germany, even Russia.

* Engineering graduates leave the field. The tendency of engineers or people educated in engineering to go on to non-related fields, such as law, medicine, business, and so forth, is an increasing trend.

* There is one that I think is the biggest and the most fundamental trend. I believe that universities are becoming more and more unattractive as career places for students and for faculty.

While I don't think there's any need to belabor that point, belabor it I will.

It's a complicated problem. Every time I discuss it with anyone we always move away from the point, because there are other aspects to the problem. I appreciate that there are other aspects, very important aspects, of the problem. However, I don't know how to solve all problems at the same time. All I know how to do is to focus on what I consider to be the root problem, and I will keep coming back in the course of this presentation to what I consider to be the root problem. The root problem is a short-term problem and at the same time a most basic long-term problem. Simply put, it is the unattractiveness of an academic career to both students and faculty. What we need for engineering education has been stated succintly. We need more graduate students, and we need more faculty members. The statistics are alarming. You're aware of them, but let me repeat them nonetheless.

* Undergraduate enrollment up in the decade 47%.

* Professional degrees and master's degrees up in the decade only 11%.

* Doctorates decreased by 24%.

* 35% of the doctorate degrees were awarded to non-U.S. citizens.

* 2,000 of 25,000 engineering teacher jobs unfilled in 1980.

There is a great concern about numbers, but there's a great concern also, about quality. People in your various institutions will say that we better lower the quality bars. They will rationalize this in order to take care of the quantities that must be obtained. Reduce the requirements! Reduce the quality. The trouble is that it will happen insidiously. No one is going to get up on a Monday morning and say, "Let's cut our quality by 50%." But every little decision will get made a helf percent at a time and over a period of five years we'll find perhaps (unless you are very, very careful) that you have in your institution compromised quality beyond repair.

Who ought to be concerned about all of this? It seems obvious that really everybody ought to be concerned. Universities have to be concerned. Engineering schools, in particular, have to be concerned. Industry has to be concerned. But you know, in listening to discussions in the past few days, I get the impression that there's a feeling on the part of some in academe that because industry is so dependent on engineering schools to graduate good engineers, it is going to come to the fore and solve the problem for its own benefit. I think the surest way to disaster is to have the smug feeling that because in the long run industry will suffer that industry in fact is going to step to the fore to solve the problem, to take the initiative.

The same thing for government. We've gotten used to the government taking the initiative ever since World War II. But I've got to tell you that, given the present administration's philosophy, another sure way to disaster is to have the smug complacency that government is going to have the foresight to come in to take the initiative and pull engineering education out of the hole.

Do I think that industry and government will help? The answers are that they not only will, but they should, and they must. I'm not talking about their willingness to put shoulders to the wheel, etc. I'm talking about initiative.

How did the problem reach this crisis proportion that we seem to be in? Let me give some possibilities.

Let me point out that the only reason I do this is not to find out who shot John, but to help reduce the chances that future crises will occur. I think the Vietnam situation and the changing social perception that emerged in our society may have had a lot to do with it. I think a lot of college graduates who at one time thought that a faculty appointment was the highest reward have come to decide that the so-called privileges and immunities of the professorial life were really not worth the differential in salary and in opportunity. I think that tenure, once the brass ring of an academic career, in many cases, has become an inflexible barrier to reasonable career paths. Put yourself in the position of one getting an undergraduate degree, who now has to weigh getting a good industrial job against the other possibility of four years or so of graduate work and then six years or so of faculty appointment before he gets put before a tenure committee which may not like the color of his eyes. That's ten years, give or take, of uncertainty with respect to his career. That is ten years that he might better spend in establishing his career on a surer path.

Our government support of universities (through originally ONR, and later NSF) after World War II, really played a major role in the emergence of this country as a major technological country. I believe that we have become over-dependent on this type of funding of both faculty and student support. I think our uncritical expectation that these funds will continue in ever increasing amounts is a snare and a delusion. I think the present administration is not going to support this expectation at all.

As a result, I would guess that little thought has been given to prudent management, and few preparations are being made for possible reductions in grants and scholarships at least until we face the crisis.

I think that inflation has played a big role in generating the problem. As a young graduate student you're in a poverty condition at the very time when an individual ought to be thinking about getting married and a couple thinking about starting a family. Even without very fancy ambitions, it becomes almost impossible in terms of finding adequate housing and in terms of starting a household. Not only that, but inflation has in fact contributed to the increasing differential between academic and industrial salaries. The inevitable result is that both students and faculty leave.

The preceding addresses the problem of non-competitive salaries and non-competitive working conditions at colleges and universities. I work closely with two colleges. In the ten years of working very closely with these colleges, I have never once seen a salary comparison with industry. I've seen salary comparisons with other educational institutions, but I have never once seen a salary comparison with industry. In trying to prepare for this presentation, I wanted to size the problem. What is the salary problem? What is the differential between academe and industry? It is next to impossible to look at the data at hand and derive a meaningful answer. The data doesn't really exist. We don't gather the data. If industry is a major competitor for your intellectual capital, your people, you had better know what the competition is doing. You had better size your people against industry. You had better understand what the gap is. Your competition is not merely other educational insititutions. In the main, it is also, very heavily, industry.

At this point, I know your heart is saying, "Yes, but Jerry, there are other problems as well." There is the equipment problem (and God knows there's a terrible equipment problem). It has been sized by the AAES in a report saying that replacing all the equipment twenty-five years old is about a forty million dollar problem. I happen to think that the equipment problem is more than a simple forty million dollar problem. If academe were to get even eighty million dollars and told to solve the equipment problem, you would again have the equipment problem in five years. There is something implicit in the structure of academe relative to technology as it exists and as it's growing that says you have to operate differently with respect to equipment. You can't operate today and the next decade with regard to equipment for pedagogical purposes or equipment for research purposes the way you have since the end of World War II. I don't have any glib answers for how you have to change, but, let me tell you, you can't do it. Money won't solve the problem. A thirty or forty million dollar or eighty million dollar grant won't solve the problem. It will ameliorate the problem. It will help you in the near term. It won't solve the structural problem.

As I said, it's almost impossible to get meaningful, consistent data with respect to wage and salary comparisons. At best guess is that somewhere between six and ten thousand dollars per faculty member on the average is going to be necessary in order to restore some semblance of parity between academe and industry. If you assume twenty-five thousand teachers of engineering, instructors on up to full professors, that says that the bill could be as high as a quarter of a billion dollars, $250 million. This figure assumes that it's going to be possible to raise engineering faculty salaries alone, without raising the salaries of science professors, humanities professors, liberal arts professors. We know we can do it. We have already done it. No medical school pays its faculty on the same scale that it pays its engineering faculty or liberal arts faculty. No law school does it. In order to get the right people you may have to pay a different scale from the normal faculty scale. I know that state-supported schools have a different set of conditions than private schools and that universities have a different set of conditions than institutes of technology. However, this problem can and will be stepped up to as a self-preservation measure.

Is a quarter of a million dollars a year a correct sizing of the problem? Paul Gray says he doesn't know whether two hundred and fifty million dollars is correct or not, but at M.I.T. they say to it that assistant professors got a twenty-five percent increase in salary. He commented that wages can't be short-term solved, but that the twenty-five percent increase had to be built into the base, and in order to do that last year M.I.T. raised its tuition a whopping nineteen percent. The motivation was quality and survival.

Getting back to the sizing of the problem, is it a two hundred and fifty million dollar problem? Ed David, president of Exxon Research, claims that's much too high a figure. His analysis of the problem (and I'll leave it to him to justify) is that of the twenty-five thousand faculty positions in engineering, only up to ten thousand represent people doing research or who are in the graduate faculty and who have graduate students. Since those are the people you should worry about, the problem instead of being two hundred and fifty million is really closer to forty, fifty, or sixty million.

Well, I've got to agree, that if in fact the problem is a fifty million dollar problem, it's almost a non-problem. Fifty million dollars a year for American Academe to raise as part of an increase in contributions from industry is a relatively easy thing to do. Two hundred and fifty million dollars I predict will not happen. It won't happen without tax credits or some other motivating elements to give industry a quid pro quo in order to contribute this amount. If in fact the problem is half a billion dollars (if you include not only salaries and wages, but other working conditions and equipment and the like) then there isn't any way, even with tax credits that you're going to get that kind of money from industry. That will mean the government very much has to be a partner in whatever the solution is.

Let me come back to the reason for talking about wages. Wages are a prime element in the attractiveness of a career. A prime element, not the only element, but a prime element. Let's see what's happened to wages for entry-level and senior positions in academe. To the best of our ability in order to make a comparison, we looked at entry-level salaries in academe and in industry for people two years from the B.S. degree. Also, we looked at salaries for people thirty-five years from the B.S. degree in both academe and industry. We looked at those figures for 1968-69, and we looked at the best figures we could get for both cases for 1979-80.

The figures showed that for associate and full professors you paid something like ten percent less for the lower quartile, fourteen percent less for the upper quartile and seventeen precent less in the mean than industry did in 1968-69 for comparable professionals. In 1979-80, the figures were twenty-three percent for the lower quartile, twenty-three percent for the upper decile and twenty-three percent for the mean. So that in ten years your top people, your thirty-five year people, went from seventeen percent below industry on the average to twenty-three percent on the average - not a very significant change.

Now let's look at the two year from B.S. in industry vs. entry level (instructor) in academe. There in the lower quartile you paid eighteen percent below industry in 1968-69. Strangely enough, at the same time, you were paying your upper decile twelve percent better than industry and on average in '68-'69 your entry level person was getting one percent less than industry was paying. Today, you pay your lower quartile seventeen percent less than industry, but you pay your upper decile twenty-six percent less than industry. This is an increase of thirty-eight percent, so that the mean entry level salary for new instructor on your faculty is seventeen percent less. That's a hell of a change from ten years ago.

You went from essential parity with industry for entry-level positions ten years ago to better than seventeen percent less than industry today. At that time of life we're not talking of seventeen percent on top of a wage which allows you to buy a camper, have a home, two dogs and a wife and three kids. We're talking about a salary level which is almost at the poverty level to begin with. So seventeen percent less is very, very meaningful. The precise numbers can be either right or wrong, but the important thing is not whether it's seventeen percent or fifteen percent or twenty percent. It's just a very big change, and it's a change in the wrong direction.

For engineering it's a truism to say it's a changing profession. It certainly isn't practiced the same way today as it was practiced ten years ago or twenty-five years ago, and the change is accelerating today. In the discussion yesterday, we were at times talking at cross purposes. Technology (for instance the microelectronic technology that Lou Rader was talking about yesterday) not only changes the manner in which you teach, it's going to change the manner in which you do the process of professional work. For example, in chemical engineering you're not just going to put in microprocessors to control your present day process. The process itself is going to change. The point is that in printers we just haven't put the new electronics on old printers. We've changed the fundamental manner in which we do printing. Because of this ten-to-one difference in electronics cost all of a sudden what was, ten years ago, a roomful of electronics can now be put in your ear. Similarly, teaching is going to change. Fundamentally, the practice of engineering is going to change. Fundamentally, the technology is going to change.

In this regard not everybody is going to go in the same direction at the same time. The point I made yesterday is that not every company is an AT&T, or an IBM, or an Exxon or DuPont. There are small companies. There are low-technology companies. Most engineers, either by choice or by fate, are going to work for low-technology companies as compared to high-technology companies. A very significant thing said yesterday was that there's something like two or three thousand people in this country who are literally designing microelectronics at the semiconductor level. That number (whether it should be three thousand, or two thousand, or four thousand) is not very big when you consider the total number of engineers. But the impact of what they do on the rest of the engineers, the rest of that million people out there is what is important and hard to divine.

The thing about the high-technology kind of engineering is that it's become very capital intensive. I wish I had some wisdom I could give you that would be practical with respect to what to do about that. But, I really believe that that is a subject for a different kind of study. A study with a different kind of introspection on the part of academe as to how small colleges and large colleges are going to take care of that situation. I fully expect that the twenty, or thirty, or maybe even thirty-five or forty leading technical institutes will take care of themselves. Cornell has its submicron facility. M.I.T. and Stanford have their large-scale integration things going. Caltech has a thing going too. But just like every college after World War II did not have its own wind tunnel, every college is just simply not going to have its own semiconductor operations. Every college is going to have to learn how to operate in spite of that. The smaller colleges, the colleges let's say, that are second line (not second class) and who graduate the bulk of the engineers that support American industry are going to have to find ways of solving that equipment problem, capital intensive though it is.

In preparation for this session, I did talk to people at the Council for Financial Aid to Education. The numbers that they have (the figures for 1980-81 aren't there yet) for 1979-1980 represented a 25% increase in the industrial contributions to higher education over the year before. I've got a simple question to ask. "Do you people think that industry is going to spend 25% more every year in its contributions to higher education?" I don't. I think they will give more, but I don't think that they are going to give 25% more every year. So if industry will not, the increase is not going to be enough to buy us a solution.

I think that you can encourage graduate study, and you can develop programs of support for graduate students to take them away from the proverty level. I think you can increase the support of cooperative research at universities.

I think you're going to have to take another look at your patent policies. Let me tell you. The studies that I've seen made about patents and the income that colleges and universities get from patents almost invariably reduce the ante. You don't make money from a patent. There are exceptions. Every once in a while there's a patent, like Warfarin, where you make a pocketful of moeny. However, the way to handle that situation is to say that the run-of-the-mill patent is near to worthless that it doesn't pay us arguing about it. So let's not argue about it. But, let's say that if all of a sudden something is done in conjunction with industry that goes beyond some trigger point, where the money is big, etc., then we want to share in it. Now I'm with you, because now you are arguing about something of substance. You're not dotting i's and crossing t's and letting your director of development or research dictate a bunch of very impractical things relative to a whole bunch of relationships that you could be having with industry if not for these outmoded and very worthless notions about who should own what, why and so forth, when the net result is practically nothing. So work out a deal that enables you to share in the big ones, and the hell with the small ones. You're not going to lose anything but you'll gain.

I think public opinion has to changed about the need and the value of engineering education at all levels. And I think that the AAES and the professional societies should be your partners in getting that kind of work done.

In conclusion, I say the crisis can be solved, but we have to stop our hand wringing and moaning. We'ver got to stop discussing ad nauseam the difficulties of the problem. We've got to remember that something in the present scheme of things has caused the problem, and there is no going back. Therefore, something in the structural scheme of things has to change, or else we'll never get out of the problem. We have to act to solve the problem in small steps if need be for there's no one solution that's going to solve the whole thing. Above all when graduate study and academic careers become desirable they will attract students and faculty.

HOW TO INCREASE THE NUMBER OF Ph.D. CANDIDATES
AND THE SUPPLY OF NEW FACULTY MEMBERS

by

Donald D. Glower
Dean of Engineering
Ohio State University

THE PROBLEM

There exists a crisis in engineering education today since there is a serious shortage of PhD graduates who are qualified for positions on our university engineering faculties. Coupled with this problem, and a part of the cause of the shortage, is the relatively large amount of antiquated laboratory equipment now at the universities.

The result of the existence of the problem:

1. Serious loss of U.S. productivity gain/world competitive positions

2. Shortage of faculty with first hand experience/know-how of the American economic system and hence the loss of the opportunity to - as a supporting agenda - educate about the American system

3. Relative loss of technical resource base necessary in the event of a national emergency

The solution: - better yet - one element of the solution - is a program designed by a committee of NSPE -
A program designed to provide 1000 new PhD graduates qualified to assume teaching positions and/or leadership positions for creative design and new technology development and its application to the betterment of our competitive position in world markets as well as in U.S. markets.

The Program Costs

Year	# of Students	Stipend	Grant	Total per Year
1	1300	$11,500	$11,500	$29,900,000
2	1300	11,500	11,500	29,900,000
	1200	13,500	11,500	30,000,000
				$59,900,000

Year	# of Students	Stipend	Grant	Total per Year
3	1300	11,500	11,500	29,900,000
	1200	13,500	11,500	30,000,000
	1100	15,500	11,500	29,700,000
				$89,600,000

Justification for program magnitude - i.e. 1000 Ph.D. degrees

"Economic growth, technological innovation, scientific research ---these are the components of progress. These are the engines that drive a country forward." [1]

25% of slowdown in productivity is attributed to reduced technological progress. [2]

"The most important source of productivity growth is the application of a new technology to the production of goods and services. More than half of the net productivity growth during 1948-1977 is attributable to technological advances." [3]

In other words, we're slipping because our generation of and application of new technology is slipping relative to our international competition of and application of new technology.

Productivity advantages/superiority is/was the cornerstone of U.S. as the land of opportunity - known throughout the world.

During the last adacemic year there were perhaps 2000 vacant faculty positions - for which $ were budgeted but for which qualified candidates were not found.

Willand Ariel Durant point out in their The Lessons of History that "Education is the transmission of civilization -- civilization is not inherited; it has to be learned and earned by each generation anew; if the transmission should be interrupted for one century, civilization would die and we should be savages again."

The engineering educator is the transmitter of technology from one generation to the next - and a partner - directly and indirectly in the generation of and the application of new technology. Yet the number of engineering Ph.D.'s granted annually, the source of our faculty, is declining significantly.

Considering the current shortage of faculty plus the number of unfilled faculty positions - there is a need for at least 1000 to 1200 additional Ph.D. degrees annually.

Further, since they will be "transmitting technology" to the next generation, they must know and understand the American way - the

---

[1] Herbert K. Meyer, editor, Fortune magazine; [2] John W. Kendrick, Economist, George Washington Univ.; [3] Frank Batten, Chairman, NY Stock Exchange

the free enterprise system - they must have experienced it - if they are to apply as well as generate new technology. Hence the need for U.S. nationals or their equivalent.

The additional Ph.D.'s are needed to teach the expanded classes of undergraduate engineers who are needed in our technical society. Consider for example the relative engineer density of the USA and Japan. We would need to double our output of BS degrees to stay even with Japan's engineering input into their technical society.

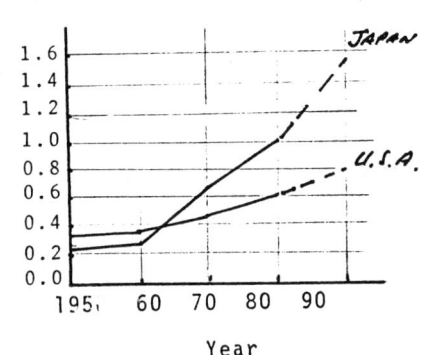

% of Total Population

[Graph: % of Total Population vs Year (1950, 60, 70, 80, 90), showing JAPAN and U.S.A. curves, y-axis from 0.0 to 1.6]

Year

Justification for $ magnitude of program:

According to Dean Ernie Gloyna of the University of Texas, the cost to produce a degree in Texas is

BS = $20,000 ÷ 4 = $5,000/year

MS = $42,000 ÷ 1 = $4,000/year

PhD = $182,000 ÷ 3 = $61,000/year

The $61,000/year for a PhD student is approximately due to the following:

	A		B
University expenses (non engr)	$15,000	+	$25,000
Engr Faculty Salaries	10,000	-	12,000
" Staff	3,000	-	2,000
" Operating	8,000	-	6,000
" Equipment	15,000	-	5,000
Capital Buildings	10,000	-	12,000
	$61,000		$61,000

Income to cover these expenses are approximately:

	A	B
Tuition	$3,000 - 10,000	
State subsidy	5,000 - 6,000	
Capital appropriation	10,000	
Equipment	?	
Remaining	30,000 - 40,000	

Sources of $'s

According to Kendrick a 10% improvement in new technology development and its application represents approximately a $3 billion/year increase in GNP due to productivity change. A 10% profit in these $'s represents $300 million/year.

Graduation of state of the art engineers could quickly impact productivity to cause a 10% change since 10% of the engineering work force turns over every year.

The current 1,320,000 engineers in the USA include ~½ working in management and "other" non-engineering jobs hence ~66,000 are in jobs where technology is applied and developed - i.e. could utilize state of the art technology - Universities send to these ranks ~66,000/year hence - work force rolls over every 10 years so up to date grads can make a difference and affect Technology Transfer.

Much of private industry has recruitment programs and spends 50-80 thousand $ per new hire. A new hire - new or replacement - is made if there is ~2½ times the annual salary in new business/lost business. Hence the output of our engineering schools - according to this model is

```
55,000 BSc Degrees @ $22,000 ea. = $3025.0 million
17,000 MSc    "     @ $25,000 ea. = $1062.5
 2,500 PhD    "     @ $32,000 ea. = $ 200.0
                                    $4287.5 million/year
```

Therefore, the Kays/Baldwin approach of asking private industry to volunteer for each hire:

```
BS = $3000 ea. ----------$165. million
MS = $4000 ea. ----------  68.
PhD = $5000 ea. ----------  12.5
                           $245.5 million/year
```

"?"

Now the first draft program proposed funding according to

	Gov't	Industry	University	Total
1st yr	$10 million	$5.5 million	$5.0 million	$20.5 million
2nd yr	$16 million	$12 million	$8.5 million	$36.5 million
3rd yr	$20 million	$20 million	$15.0 million	$55.0 million
Steady State	$20 million	$20 million	$15.0 million	$55.0 million

where $3000 per student per year was provided to the student's department.

A meeting in St. Louis of Deans and Industry representatives made the following points:

1. The Federal government should fund the whole program

2. The $3000 cost allowance was inadequate and had to be increased to match the stipend if the private schools were to be able to compete

3. The university and private industry would not contribute -- we'd have to wait years -- for modest success....

Hence the evolution to the following program which was what would gain the support of all

Now what happens if we do not react to the Crisis? We make veiled threats inferring catastrophe....

Actually nothing dramatic will happen - the manpower balance will take care of itself - and no one will notice any change -

There is no doubt that in time we'll shift our programs to include the state of the art technology.

There is no doubt that we will fill our faculty vacancies - BUT - we'll have a different country/society - depending on whether or not we do our job - as the transmitters of technology to the next generation!!

So let's get behind this program which has received wide approval - it is acceptable to our various constituent groups - the private schools, the public schools and private industry.

Now where do we get the $'s?

(1) The Federal Government

    a. a 10% improvement in the rate of productivity gain represents $3 billion per year or about $300 million in corporate profits. Can we share these $'s? I think we can.

    b. a year saved in the trade wave to gain proper balance represents $'s in the $100 billion range. Can we share in these $'s? I think we can.

We can pool our political forces and get federal legislation to provide these dollars.

(2) Private Industry and University as a team

    a. The Kays/Baldwin model will produce significant $'s from private industry - perhaps $200-300 million per year. Can this be done? Yes, but this is only a partial solution - but it will help!!

b. Universities are today contributing about $40,000 per PhD degree - or about $100 million per year. We can use this as matching money to either private industry or government.

I ask that this workshop endorse and help implement the proposed program -- in the action/conclusions of the Conference -- get behind the program - let's stop talking - let's act!!

In conclusion, I want to re-emphasize -

The important concept - the facts - what we must impress upon our colleagues is the following:

1. We will fill our faculty ranks with individuals who are excellent in intelligence and ability, HOWEVER many will have their formative years in a foreign land - under a system foreign to the American free enterprise system of economics - and they will rise to leadership position on our faculties and they will greatly influence our future engineering education and our technical society.

2. The time span to meet the marketplace challenges of Japan and other technical societies can be equated to dollars and to our way of life.

    a. If we educators do nothing the US will win the technical - the international trade war - in perhaps 10 or 15 years BUT we'll be a different country due to our expenditure of $'s on priority items such as Petro, etc. and Japanese/German products - and a balance of trade deficit.

    b. If we develop an action plan to convert our education programs - to integrate throughout the programs - state of the art technology - and we graduate youngsters who are state of the art - if we transmit state of the art technology (to paraphrase Durant) - we can shorten that 10-15 years to 5-10 years and we'll have a country different from that which would otherwise be - we'll still be great - BUT - each year saved represents billions of $'s to the American economy - this is what we must preach - SELL - WE CAN MAKE A DIFFERENCE

Get behind the NSPE/ASEE/deans approved program --

Let's have ACTION - not just words -

"ACTA NON VERBA!"

IMPLICATIONS OF INCREASING PERCENTAGE OF FOREIGN
NATIONALS ENROLLED IN GRADUATE PROGRAMS FOR
ENGINEERING

by

W. Robert Marshall
College of Engineering
University of Wisconsin - Madison

The implications of an increasing percentage of foreign nationals in graduate engineering education would appear to be not too difficult to predict for different scenarios. We need only look at our experience of the 70's and then do some careful extrapolating based on certain assumptions for each scenario. The scenarios which I believe may be in the realm of possibility are:

1. Enrollments of U.S. citizens in engineering graduate programs continue to decline while faculty research continues to grow. (Assumption: industrial demand and salaries for B.S. graduates continue to increase at a significantly higher rate than graduate student stipends). This would tend to increase foreign student enrollments even higher.

2. Enrollments of U.S. citizens in engineering graduate programs level off and begin to increase for whatever reasons. (Will there be a significant increase in women graduate students?). An unlikely scenario. This could result in a decrease in foreign student percentages.

3. Engineering research in universities receives unusual support and encouragement from government and industry creating a greater demand for graduate students.

4. Engineering research in universities experiences a serious decline reducing the demand for graduate students.

A basic assumption to all of the above is that undergraduate enrollments remain very high even though controlled.

There could be other scenarios caused by new tax laws.

Fundamental to our considerations is the role of the graduate student in engineering education. The graduate students have three basic roles:

1. They seek advanced degrees.

2. They conduct research - usually as part of a professor's program and as part of the degree requirement.

3. They assist in undergraduate instruction.

The latter two roles are, of course, vital to a college's research and instructional programs, even though the first role may be of primary importance to the student.

In most graduate engineering programs, foreign students have participated in all three roles to varying degrees. However, their participation has increased dramatically over the last five years as the numbers of U.S. students engaged in graduate study have decreased alarmingly.

Reasons for the presence of foreign graduate students:

1. Individual desire to study in the United States and family pressures.

2. Foreign programs in the college.

3. U.S. government programs.

4. Foreign government programs.

5. Faculty interests in foreign countries.

It is impossible to present an exhaustive review and survey of the foreign student situation in all colleges of engineering. I have, therefore, elected to use the experiences in our College of Engineering. University of Wisconsin - Madison. The University has long been recognized as one which encourages and promotes foreign or international programs. As a consequence we have large numbers of foreign students in our graduate programs with the largest percentages being in the College of Engineering. Our Graduate School closely monitors the foreign student applications and permits for admission in all colleges but it has taken a special interest in the situation in the College of Engineering; therefore, we have fairly precise trends for the past several years. The experience in our college is probably not greatly different from that of other research universities. Our percentage of foreign students has been steadily increasing even though the total numbers of graduate students has remained somewhat constant. The situation in 1978 showed 703 post-graduate students, 43% of them were non-U.S. students. In 1979, 717 -- 42%; in 1980, 720 -- 45% foreign. Foreign enrollments have gone up fairly steadily since 1965 with an interesting peak showing in 1971 when we had 45% foreign students in the College. Among eight degree granting departments, the department which currently has the highest percentage of foreign students is the Department of Mechanical Engineering with around 62%; the lowest percentage, 25%, is in Nuclear Engineering, all other departments range from about 30% to 60%.

With respect to the national picture on the percentage of foreign graduate students in engineering, the data for enrollment in the 79-80 period was compiled and presented by both the National Science Foundation and the Engineering Manpower Commission. The data presented at that time showed for all institutions 39,282 full-time graduate students of which 16,179 were foreign, equal to 41.2%. Among the ten leading institutions in terms of doctoral outputs the foreign student percentages was 39% and ranged from a low of 31.2% at MIT to a high of 54.1% at Ohio State University. All other institutions showed 42% foreign student enrollment. When one looks at the percentage of foreign students getting doctoral degrees during that time, approximately 32.6% were foreign. This percentage, of course, has increased in the past two years, even though the total doctoral output may not have changed significantly.

As I indicated earlier our graduate school has been monitoring graduate school applications in engineering for several years. It has compiled data as of June 27, 1981 as follows: of the total applicants in engineering and certain related sciences, there were 1,034, of which 605 were foreign. There are 142 pending, of which 105 are foreign and permits have been issued to 473, of which 288 are foreign or 61% foreign student permit issuance. In the physical sciences division total: there have been 905 permits issued, of which 447 are foreign or 49%. For the Graduate School as a whole, including all fields, 3,419 permits have been issued, of which 948 are foreign or 28%. A noticeable trend is developing in other fields than engineering; physical sciences are showing significantly increased numbers of foreign students in areas such as mathematics, computer sciences, and physics. The foreign student permits for engineering have been tracked by the Graduate Schools since November 1980, and at no time during these months was the percentage lower than 50%, except in one case where it was 49% and it ranged to 62% and 61% in May and June. Clearly, the numbers of foreign students are becoming a major factor and fraction of our graduate program, and the presence of so many students from other countries clearly poses many problems. Many problems with which we are already familiar but now are growing in magnitude because of numbers.

With a large number of foreign students to educate, support, house and look after their personal needs, there is much time consumed byfaculty and advisors who are responsible for foreign students. In our University we have a special office for foreign students to take care of their personal problems; e.g., to handle health needs, immigration needs, etc. Within the university community we have groups of faculty and volunteers who have bery strong concerns and interests in students from other countries, especially those who come from warm climates and arrive in Madison, Wisconsin when the temperature is -25°F. As a matter of fact, this has led to the formation of an Overcoat Club. Students from India, Indonesia and those countries which are always in warm, temperate climates, experience a severe shock when they arrive in Madison in the dead of winter. These unexpected environmental conditions create problems for the students and we obviously cannot ignore these personal needs, so an impact of increased numbers of foreign students is reflected in increased requirements to five them attention and to provide them with assistance on personal matters; this is all aside from their educational needs and goals.

Clearly, one of the major concerns which we have for all the students from other countries who will have English as a second language will be the problem of communication. The need for foreign students to speak reasonably good English is important primarily for completing their courses successfully, for writing theses and for taking oral exams. The need becomes critical when it is necessary to employ foreign students as teaching assistants. This can be a major impact on our engineering educational programs. It is impossible in our College of Engineering to do all the required teaching without hiring TAs who are from another country and who speak English as a second language. It is sometimes a difficult problem in departments which are experiencing high enrollments and large classes to obtain enough TAs who can communicate adequately with students.

It is not a surprise to anyone in this audience that the reason for the increased percentage in foreign graduate students in engineering is due to the decreasing number of U.S. students. This is probably a more important problem for consideration at the moment than the impact of increased foreign students. I think the problem the colleges may face in the future if they're going to avoid a crisis in teaching and meeting the growing demands for engineers will be whether or not we will be forced to hire ever increasing numbers of non-U.S. faculty members to fill our vacant positions. There seems to be a small trend in this direction in our College at the present time. We currently have about 10% of our faculty of non-U.S. origins. We do not, however, sacrifice professional and technical quality to hire a replacement or to fill a vacancy. We have been fortunate in being able to attract extremely competent foreign Ph.D.s educated in the U.S. who are here on an immigration visa. It would be an interesting statistic to develop to find out if other colleges are experiencing or intend to experience an increase in the number of non-U.S. Ph.D.s on the faculty. Such hiring might be done on a temporary or interim basis. If we learn that the non-U.S. Ph.D.s may be just as good as or better than some of our W.S. Ph.D.s, then we face the possibility that sometime in the future the composition of our faculty may be much more cosmopolitan than it is today.

I believe we should ask some pertinent and hard questions regarding the "problem" of foreign graduate students. Is it a singularly serious problem to have the numbers of foreign students increase as high as 60% of the graduate students if the foreign students are well qualified, if they are capable of conducting good research and if they are capable of communicating adequately in English. It behooves us I believe to look at this problem with a fair amount of realism: first, can we expect a change in this trend within the next eight or nine years; second, what factors will cause a reversal to take place, third, are we not, as a matter of policy in many colleges encouraging foreign graduate student enrollments in engineering as a result of our various foreign programs which automatically generate a flow of students to our universities? I am sure there is not a uniform policy among colleges of engineering with respect to engaging in foreign programs. We have a unique situation at our college at Madison. We entered into foreign program operations as early as 1953 when aid to underdeveloped countries was in its infancy. Our faculty appeared not only to enjoy the experience of foreign activities but to support it and encourage it. As a result, we have had involvement in countries all over the world over the past three decades. Our faculty have individually spent much time visiting in foreign countries. This has automatically generated a flow of foreign students to our college as a result of both personal interactions among our faculty and foreign faculty, as well as through formal programs which the college has contracted to operate, such as those in India, Singapore, Indonesia, Mexico, etc. This situation is true throughout our university and I believe a great deal of pride is taken in the foreign programs that have operated in most of our colleges and schools. The question can always be raised, what are the values of these programs? how do they help our educational processes? what do they do for our students? and I believe these questions must always be asked. Our experience is that the benefits are both short-and long-range. They are short-range when our students can be active participants and benefit from the exposure to other cultures and broaden their horizons during their academic careers. The benefits are long-range when our programs are designed to educate faculty from developing countries and when we must wait the passage of time to determine whether faculty so educated have transfixed these benefits to their own country. Sometimes, there are unexpected long-range benefits from the education we provide foreign students who return to their countries and rise to positions of prominence. We live in a world that is small, in terms of rapid communications and travel. Peaceful interaction with other countries and nations is essential, and can be achieved, in part, from the U.S. education of students from all parts of the world. Education and understanding are major forces in easing tensions that seems to exist among nations. However, we must be realistic about how much we can do with the resources we have and I believe this is the real problem facing university administrators and engineering college deans when foreign programs are considered. The question will be raised: are we putting more resources than is appropriate in the support of students from other countries? how can we properly assess what this support should be? It is generally a "given" that undergraduate foreign students are a financial asset, they pay their full tuition and full cost of education in most instances. Graduate students may or may not come with their own funds. Generally, we must find support in one form or another; this becomes a serious problem. Finally, we must decide the question posed earlier, namely, can we afford and is it good policy to increase the percentage of non-U.S. faculty in our colleges of engineering? what problems does this pose for us in the future? what recommendations should be made, if any, on this very difficult question. We are trying to find faculty from a very limited number of Ph.D.s in engineering, we are seeking desperately to find them becuase we have many students to teach at the undergraduate level. If we cannot find them and if we decide to limit the number of foreign faculty, then we must begin a tighter enrollment control and continue to deny access to U.S. students who want to obtain an engineering degree and practice this very exciting profession.

Limiting enrollment at the present time does not seem to be a very sound policy throughout the country when the need for engineers appears to be continuing strong and many suspect it will remain strong and possibly increase if our economy begins to turn around. All these problems, in my judgement, are worthy of discussion and consideration at this conference and I hope that there will be innovative suggestions made.

EFFECT OF FEDERAL SUPPORT FOR PROGRAMS AND STUDENTS ON GENERATION
OF NEW FACULTY

by

Ross J. Martin
Associate Dean and Director
College of Engineering/Engineering Experiment Station
University of Illinois at Urbana-Champaign
1308 West Green Street
Urbana, Illinois 61801

We are all acutely aware of the current crisis in engineering education. Colleges of engineering are particularly hard hit in the areas of faculty shortages, faculty salaries and lack of modern plant and facilities to provide appropriate laboratory experience for our undergraduate and graduate students in current engineering topics. It is true that variations in number of students and numbers of degrees granted at all levels make projections into the future difficult.

However, the trend in engineering baccalaureate degrees has shown a steady growth since the late 1930s. Also the demand and production of Ph.D.s in engineering increased by seven-fold during the period 1950 to 1972 when the production of Ph.D. candidates leveled off and decreased slightly: but the demand for Ph.D.s continued to increase at a steady rate -- creating our present shortage of Ph.D.s for all career options including those entering professorial positions. Thus it is not difficult to predict that for at least the next 5 to 10 years, the faculty-shortage "crisis" in engineering will be seriously impacting our colleges of engineering.

The current shortage of qualified (Ph.D.) faculty has been estimated at from 1800 to 2200 vacancies or roughly 10% of the 22,000 professorial faculty in our more than 250 engineering colleges with ABET accredited engineering degree curricula. This does not count the appropriate growth needed in the size of engineering faculties to relieve the approximately 30% overload which has been generated by increased undergraduate enrollments since 1973-74. Further, the current lack of competitive faculty salaries in nearly all engineering colleges is creating a tendency for senior faculty to arrange early retirement and leave academia for industrial and high-level government positions. Junior faculty are also being sought by industry and government with attractive cash salary offers as well as improved benefits-- sometimes including annual bonus and stock option opportunities. Industrial (and governmental) competition for faculty often includes an attractive campus-like atmosphere in research laboratories which are well equiped with modern instrumentation and experimental facilities and staffed with appropriate technical and clerical support personnel.

New engineering faculty have been traditionally drawn from the new Ph.D. recipients of U.S. engineering colleges. In these time current faculty shortages, the number of new Ph.D.s is at a low point and the trend is still downward. The recent engineering Ph.D. production rate of from 2500 to 2800 per year is known to be much too small to satisfy the variety of needs that this group of highly sought-after people must meet. This is particularly true when one considers that nearly 1000 (35%) of these Ph.D. degree recipients are foreign nationals and that the large majority of these foreign nationals are required or intend to return to their country of origin. It is important to note that the number of foreign national Ph.D.s graduating has remained quite constant (at about 1000 per year) and that the decrease in total Ph.D.s from a high of 3744 in 1972 to our present 2751 Ph.D.s in 1980 has been almost totally a decrease in U.S. citizen Ph.D.s. Foreign nationals who choose to remain in the United States after obtaining their Ph.D. have been "sponsored" by U.S. colleges of engineering as faculty members and a significant number of these foreign Ph.D.s are currently serving as professorial staff in our nation's colleges of engineering. More recently industry has been recruiting a larger share of these foreign nationals as a means of satisfying the growing industrial demand for Ph.D. engineers. Recruitment of foreign national Ph.D.s by both universities and industry represents an important utilization of this already critically scarce commodity -- new engineering Ph.D. graduates! Foreign nationals who stay on in this country after the Ph.D. are a valuable asset to the nation and add materially to our pool of well-trained engineering manpower. Ph.D.s educated in foreign institutions and immigrating to the U.S.A. have also provided some relief to the nation's Ph.D. manpower shortage. However, the only realistic solution to the current crisis in engineering education is a substantial increase in the number of domestic Ph.D. engineers being produced each year by our colleges of engineering.

Let us look at the various actions that can be taken to increase the Ph.D. production of our engineering colleges by enlarging the numbers of well-qualified candidates including an increased number of minorities and women in the graduate educational pipeline.

First, we are in a fortunate position related to the number and quality of our present and recent past engineering baccalaureate degree recipients. Since about 1973 or 1974, the interest in engineering as a career has increased sharply among high school graduates. Year after year engineering colleges have received more and more applications from better and better qualified students as measured by both high school percentile rank and SAT or ACT ability scores. Further, the high school study patterns of these applicants to engineering colleges across the country have indicated more high school math and science preparation. At most engineering colleges this flood of applicants has, through a selective process of admissions, provided a higher and higher quality of entering freshman and progressively enhanced the entire undergraduate engineering student body.

At Illinois, for example, during the years of decreasing interest and enrollment in engineering as a field of study (1969-73) we maintained a floor of the upper-third of the high school graduating class and ACT of 24. These students were capable of completing engineering studies through the B.S. level provided that they were motivated. However, the attrition was high and many - although they finished college - did not complete their work in engineering. As interest in engineering studies grew in the mid-seventies and our admissions level was limited by a lack of corresponding growth in resources, faculty size and facilities, our "progressive" admissions policies to maintain an entering class of approximately 1000 freshmen resulted in our present very selective admissions in which the median entering freshman student has a high school percentile rank of 95 or higher and ACT composite scores of over 28 - also representative of the top 5% of ACT scores nationally. Further, the quality of transfer students from other colleges on campus and from two- and four-year institutions has improved correspondingly by the process of progressive admissions. This quality selectively among our engineering undergraduates means that recent graduating classes, which now total nearly 60,000 per year, are among the most able of the past two decades. This is one of the most positive and encouraging aspects of our current engineering educational picture.

Industry is well aware of the high quality of these recent engineering B.S. graduating classes and has bid strongly for their services to satisfy the high demand for graduate engineers.

Unfortunately, colleges of engineering have not been especially competitive in their attempts to convince the top B.S. candidates in the nation to go on to graduate studies - at least to the master's degree, and hopefully to the Ph.D. A part of this failure is the ever-widening gap between the starting salaries offered by industry to new B.S. engineers and the level of stipends of graduate fellowships, traineeships and part-time assistantships being offered by colleges of engineering to new and continuing graduate students. A review of recent (1979-80) published stipends for fellows and one-half time graduate research and teaching assistants shows a range of from $4,500 to $7,000 for the academic year appointments. When compared with the $20,000 to $23,000 per year salary being currently offered to new B.S. engineering students, it is little wonder that the top graduating seniors have chosen not to continue as graduate students for the Ph.D. or even the M.S. degree.

Ideally for maximum utilization of the nation's engineering human resources, the upper 20% of the nation's current 60,000 engineering B.S. graduates should be continuing on to graduate school with the educational objective of obtaining a Ph.D. in the engineering field of their choice. Additional B.S. graduates from the upper half of their graduating classes should be seriously encouraged to pursue the M.S. degree in order to maximize their career contributions to the nation's engineering needs over the next several decades.

The highly educated (Ph.D.) engineering talent resulting from a dramatic increase in graduate study by our best B.S. graduates is what is needed to solve the crisis in engineering education. Further, it would provide the needed Ph.D. engineering graduates who could study and solve the ever more complex engineering problems of productivity and high technology that this nation must address for its own economic growth and to compete with other industrial nations (Japan, West Germany, U.S.S.R., etc).

How can the nation's colleges of engineering implement this necessarily large increase in graduate enrollment and the resulting increase in new Ph.D. engineers? It will not be easy and the effort will require considerable assistance from the federal government, from industry and from the colleges of engineering including resources support from the parent universities and their funding sources both public and private.

Specific program suggestions are as follows:

I. Establish a substantial number of additional fellowships, traineeships, and graduate research and teaching assistantships with starting salaries of $11,000 or more per year with step increases for second and third year graduate students. These stipends should be in conjunction with tuition and fee waivers or above any tuition and fees assessed to the student. A substantive proposal developed by a committee chaired by Dean Donald Glower seeks federal funding to provide for 1300 new fellowship-trainees each year in a three-year program yielding 1000 new Ph.D. degree candidates each year. The steady-state cost of this program is estimated at $90 million annually (in 1980 constant dollars). Additional fellowships, traineeships and graduate assistantships funded by concerned industrial corporations would also be effective in obtaining the increased graduate enrollment and Ph.D. degree growth needed in the 1980s. A few corporations have instituted special fellowship programs aimed at graduate students indicating a career objective of engineering college teaching. This action is applauded and hopefully it will be followed by other large corporations.

For this program to be successful and to provide real growth rather than a substitution for the current graduate student body (approximately 13,000 doctoral degree candidates in 1979-80), the stipends currently being paid must be increased accordingly. The resulting cost increases will affect the colleges of engineering (in the case of teaching assistants' salaries) and the federal government and other sponsors of research (in the case of research assistants' salaries). Since the stipends and fringe benefits (largely tuition and fees) and related indirect costs of the graduate research assistants constitute a substantial share of our present research expenditures, a large increase in present day research dollar volume primarily supported by federal funding will be required to maintain the support of the same number of graduate research assistants.

II. Provide "released time" from undergraduate instruction to provide for the additional load on engineering faculty to teach the formal graduate courses and research training supervision needed by the increased graduate level enrollments. The solution to this problem must, in general, come from the colleges of engineering and their supporting institutions. It is recognized that the faculty effort related to one full-time graduate student is approximately three times the effort required by a full-time undergraduate student. The ideal solution would be to employ more engineering faculty. However, this is the very shortage we are attempting to remedy. Thus in the short-run, engineering colleges must provide the means to make the efforts of the limited number of present faculty more efficient and productive. Support personnel, both technical and clerical, can increase the time available for faculty to teach, to consult with students and to maintain a viable research program. The use of more mature undergraduate and graduate students to teach and supervise less advanced undergraduate and beginning graduate students has been successful in previous faculty overload situations such as the post World War II and Korean War engineering enrollment "bulges". Hiring part-time adjunct faculty from among experienced engineers in industry and private practice is also a possibility.

One of the most time-comsuming responsibilities of engineering faculty is the supervision of thesis research and the conduct of the faculty member's own research activities necessary to maintain this supervisory capability. In recent years these faculty efforts have been provided by the universities during the academic year on a "no-charge, cost sharing" basis. In colleges of engineering with doctoral programs the faculty effort involved in research ranges from 25-50% of the professors' full-time effort. The funding of up to 25% of academic year effort of professors conducting sponsored research would compensate for the "released time" of the faculty and assist in the implementation of the faculty support package described above. Equivalent procedures were used with excellent results in the program for expanding the federally supported materials science research in the interdisciplinary Materials Research Laboratories during the 1960s.

III. Modernize the laboratory facilities used for instruction of undergraduates and beginning graduate students. Current laboratory facilities of engineering colleges are very old, expensive to maintain and inefficient to operate. Inadequate equipment and outdated instrumentation result in a wasteful use of time by both students and instructors. Most engineering colleges find it difficult or impossible to introduce new technology into their curriculum due to the lack of appropriate equipment and facilities. For example, the teaching of computer-aided design/computer-aided manufacturing in engineering colleges requires a substantial investment in computers, terminals, and software to provide students with this background which is sought by and useful to industry. President George Low of RPI estimated in 1978 that $1,500 per engineering baccaluareate degree granted is needed as an annual replacement cost of instructional laboratory equipment. In 1982 dollars this would be $2,000 or $120 million per year nation-wide for replacement costs alone. New areas of technology (CAD/CAM, lasers, semiconductor circuits, etc.) represent new equipment needs estimated at up to $500 million and would add significantly to the required maintenance-replacement costs.

IV. Provide research equipment support for the increased research activity related to the increase in doctoral level graduate students conducting thesis research. Numerous studies have been made on behalf of the National Science Foundation related to the needs for research equipment to maintain our present national basic research effort. Specialized research equipment needs are well beyond the financial resources of the average research grant or contract. Special research equipment programs have been initiated by NSF but considerably more funding is required to meet the present and future needs.

V. <u>Create an environment that will attract and maintain faculty in our colleges of engineering.</u> If we expect our present faculty to be influential as role models for future faculty while they are in the graduate training mode to obtain their Ph.D. - or even while they are still undergraduates - we must provide living and working conditions that are favorably competitive with those found in industry and government. Also, in order to attract new faculty with Ph.D.s in engineering we must provide a competitive salary and appropriate opportunities and working conditions so that they will have the best opportunity to succeed and remain in educational careers.

This requires a substantial increase in salary, the availability of a modern and efficient plant and facilities for teaching and research support staff to permit concentration on duties uniquely suited to the faculty and appropriate appreciation from fellow faculty members, administrators and the general public.

As the term "crisis" implies, the current state of engineering education is serious and is rapidly reaching a condition of instability. Positive steps to remedy the situation are urgently needed. Only with the cooperative efforts of the federal government, industry, university administrations and colleges of engineering can we hope to find a solution to this crisis in the near future.

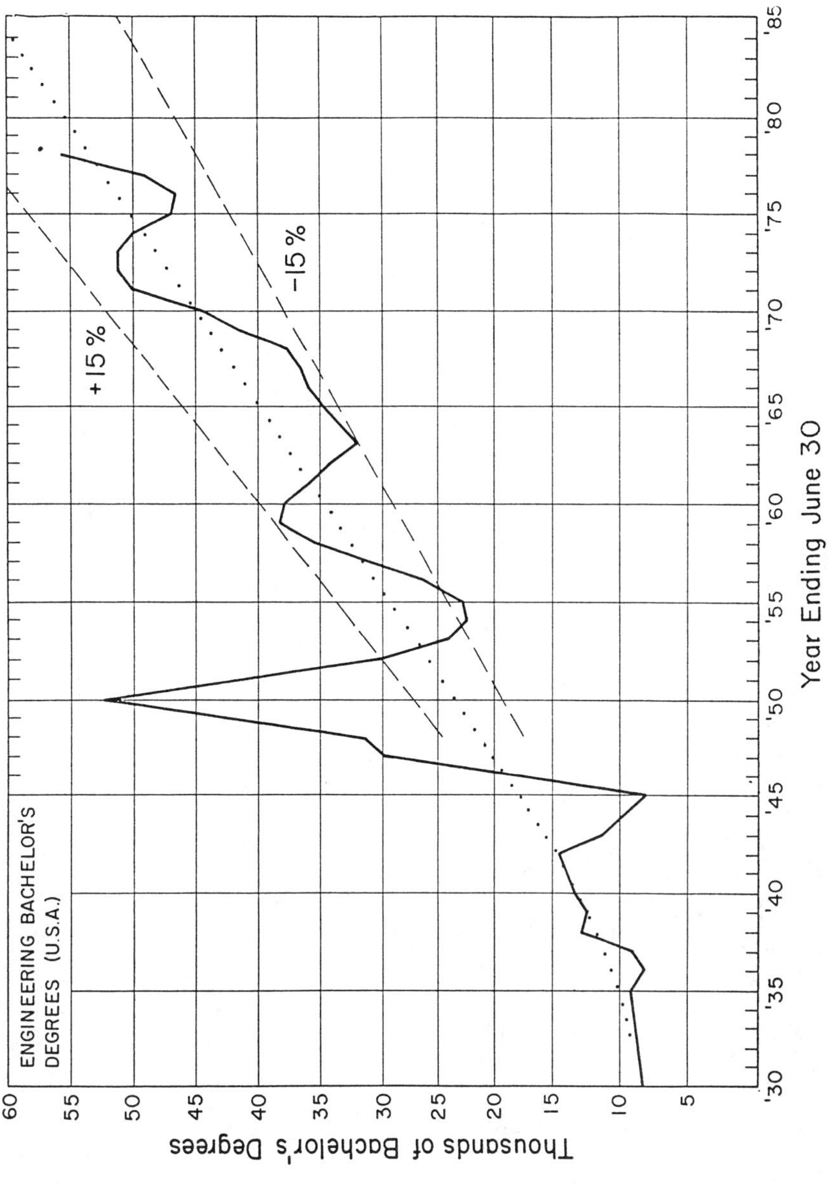
ENGINEERING BACHELOR'S DEGREES (U.S.A.)
+15 %
-15 %
Year Ending June 30
Thousands of Bachelor's Degrees

CONCEPT OF A PROFESSIONAL SCHOOL OF ENGINEERING

by

Charles H. Samson, P.E.
Acting President
Texas A&M University
College Station, Texas 77843

SUMMARY

Background is provided concerning the concept of professional schools of engineering. During recent years, considerable discussion of this concept has occurred among engineers and employers of engineers. Criteria for professional schools of engineering are discussed as developed by the National Society of Professional Engineers and by various writers.

Many questions have been raised by members of the engineering profession, including engineering educators, and by employers of engineering graduates as to the implications of the professional engineering school concept. Some of these questions are stated to show the uncertainities that exist concerning the merit of this concept.

One model of a professional school of engineering is discussed that might satisy some of the need for professional schools without threatening the effectiveness of existing educational programs. This model would require a pre-engineering educational experience of satisfactory research-oriented advanced work, and would permit the offering of a variety of degree programs by a given institution. The framework leaves open the question of the minimum formal educational requirement for entry into the engineering profession but is adaptable to future changes that might occur within the profession.

BACKGROUND

Since the early 1970s, considerable discussion has occurred concerning the need for professional schools of engineering. One of the arguments used in favor of a professional school is based on comparisons of engineering with other professions, such as medicine and law. The significance of such comparisons depends upon the importance that one attaches to common characteristics.

In an article in 1970 entitled "Professional Schools for Engineers," Lohman compared engineering education requirements with criteria for the accreditation of professional schools of medicine, law, nursing, pharmacy, architecture, dentistry, and veterinary medicine. He concluded that five characteristics are in general common to these professional schools.

1. Admissions criteria unique to the professional college.

2. Faculty members with relevant professional experience and who maintain professional competence as a duty, not a privilege.

3. Relevant intern or professional experience within the jurisdiction of professional college prior to acceptance as a full-fledged practitioner.

4. Jurisdiction over the professional curricula and professional degrees.

5. Institutional autonomy that is made possible with a separate budget.

In an article entitled "Professional Schools of Engineering"[2] published in 1972, McCollom discussed these criteria in relationship to a proposed professional school of engineering at Oklahoma State University.

In July, 1971, following a study by a special task committee, the National Society of Professional Engineers (NSPE) adopted NSPE Professional Policy No. 104.[3] This policy listed the following seven "characteristics" of professional schools of engineering:

1. Admissions criteria unique to the professional school.

2. Jurisdiction over the professional educational programs and professional engineering degrees.

3. Jurisdiction over the criteria for appointment, promotion, tenure, and activities, of the faculty.

4. Faculty members, with relevant professional experience, who maintain professional competence as a responsibility rather than as an option.

5. Supervised internship or professional experience by the graduates as a prerequisite to acceptance as full-fledged practitioners.

6. Continuing contact with professional practitions, both nationally and locally, regarding: qualifications for appointment to the professional school faculty; curriculum content; and, the type and duration of acceptable internship or professional experience required of graduates.

7. Autonomy within the institution that is made possible with a separate budget.

In his "President's Report to the Council" (ECPD),[4] Lohman provides a comparison of engineering educational programs with other selected professional programs for the years 1900, 1940 and 1972 as to structure and length. This comparison reflects a striking change in program length for other professions during this period with engineering retaining the four-year degree for entry into the profession.

While no attempt is made here to provide a complete listing of publications on the subject of professional schools of engineering, Refs. 5-17, published during the period 1972 to 1981 are representative of the active interest in the subject. Some of these favor the concept; some oppose. Various models of professional engineering schools are proposed.

In 1974, a national conference on professional school of engineering was sponsored by NSPE[18]. A conference publication was produced that contains not only material related to discussions at the conference but also background papers selected from other sources.

NSPE, through the work of various task committees, has continued its interest in professional schools. In March, 1978, it published the proceedings of a workshop on "Developing Guidelines for Recognition of Professional Schools of Engineering."[19] In January, 1980, NSPE Professional Policy No. 104 was modified to read as shown in Appendix A.[20] Appendix B provides the NSPE guidelines for recognition of professional schools of engineering presently in effect, also approved by the NSPE Board of Directors in January, 1980.[20] Since establishing these guidelines, NSPE has endeavored to find an approach by which engineering schools could be "recognized" by NSPE as "professional schools of engineering." This intent has generated much discussion in the Society. While there appears to be a genuine desire by the NSPE Board of Directors to pursue this program, there is a concern that such action action be taken in a way that would not be in conflict with the functions of ABET. It is timely to note that NSPE has requested ABET to permit NSPE representatives to participate in accreditation visits in those cases in which a school wishes to be recognized as a "professional school of engineering."

Further, at the NSPE annual meeting in July, 1981, the NSPE Professional Engineers in Education Board of Governors reaffirmed its previous statement of support for the professional school concept. The NSPE Board of Directors later in this meeting approved the recommendation that NSPE proceed in implementing the professional school after

1. ABET meets in the fall of 1981;

2. The NSPE Education Advisory Council meets in October, 1981;

3. A survey of deans of engineering schools having ABET-approved curricula to determine the extent of their interest in the professional school program. The survey would be conducted by the Professional Engineers in Education in cooperation with the NSPE Professional Schools of Engineering Recognition Committee.

THE PROFESSIONAL ENGINEERING SCHOOL CONCEPT

It seems apparent that there is no single statement of the professional school of engineering concept that has received wide acceptance in the engineering profession. For the purpose of this discussion, NSPE Professional Policy No. 104-A--"Professional Schools of Engineering" (Appendix A) and "NSPE Guidelines for Recognition of Professoinal Schools of Engineering" (Appendix B) will be used to characterize the professional engineering school concept.

QUESTIONS CONCERNING THE CONCEPT AND ITS IMPLEMENTATION

In the past ten years during discussions of the need for professional schools of engineering, many questions have been raised by engineering educators, employers of engineering graduates and the engineering profession in general. It is useful to list some of these questions:

1. Would the adoption of the concept imply that existing engineering schools are "non-professional"?

2. What does the concept imply as to length of formal engineering education? (Five years? Six years? Seven years?)

3. What degree or degrees would be awarded? What would be the status of present baccalaureate engineering degrees? What would be the status of present graduate degrees in engineering? Would new degree designations be desired? To what extent would professional schools encompass existing graduate degrees (e.g., M.S., Ph.D.)?

4. What would be the relationship between existing graduate schools and professional schools? Would new basic or advanced courses be required? Would advanced professional school courses have graduate designations and require the customary approval of university graduate schools?

5. Would admission to the "professional school" require formal "pre-engineering" academic work of a specified quality? What other special requirements would there be (e.g., writing proficiency examinations, interviews)? What other entry points to a professional school should there be? What happens to students not admitted to the professional school? What progress standards should exist?

6. Should all engineering schools become professional schools?

7. What would be the impact on industry of professional schools? Would there continue to be a pool of baccalaureate-level engineering graduates sufficient to meet the needs of industry?

8. What would be the implications of professional schools with respect to faculty qualifications? What percentage of existing engineering faculty meet the needed criteria? How much non-academic experience would be required? What importance would be attached to research experience?

9. How could advanced professional school students receive

financial support? Research projects? Fellowships? Teaching assistantships?

10. What would be the attitude of engineering faculty toward professional schools? What would be the effect on engineering faculty research activities?

11. To what extent should other professions, such as medicine and law, be used as a basis for comparision with engineering?

12. To what extent should professional schools or professional programs be "recognized" pr "accredited"? By whom should "recognition" or "accreditation" be given?

PROFESSIONAL SCHOOL MODELS

In considering the many question that can be raised, one could develop a wide variety of models depending upon the type of response to the questions (e.g., see Refs., 14 and 16). A desirable model would be one substantially satisfying the needs for professional schools without threatening the positive feature of existing engineering programs that serve the public in general, the profession, the institution, students, faculty, and employers. Does such a model exist? Certainly the diversity of attitudes that exists casts a shadow over the prospects for producing significant changes over the status quo. While there is certainly not agreement that present engineering education is all that it should be, there are so many ways that disagreements can occur regarding proposed changes that it would seems the most likely occurrence is no change at all.

No attempt will be made here to examine a number of alternatives. For the sake of discussion, however, one model will be illustrated that might merit consideration. With minor modification, Fig. 1 is taken from Refs. 15 and 21, with the exception that the block showing "4 YR B.S. DESIGNATED" appeared as "4 YR UNDESIGNATED" in the references cited. (The model has similarities to others that have been suggested - e.g., see Refs. 14 and 16.) The heavy line has been added to indicate the bounds of the "professional school." While one could attach different significance to elements within the professional school, the following features can be assumed:

1. Entry into the professional school would normally follow two years of pre-engineering college-level work of a prescribed quality, and perhaps certain proficiency examinations (writing skills, for example).

2. Entry could follow other academic work (perhaps including other degrees), but satisfying at least the equivalence of the pre-engineering block would be a requirement.

3. Following approximately two years of satisfactory work in the professional school, a B.S. degree in a designated area could be awarded. Depending upon an evaluation of performance (and possibly further examinations), this degree could terminate

ENGINEERING EDUCATION

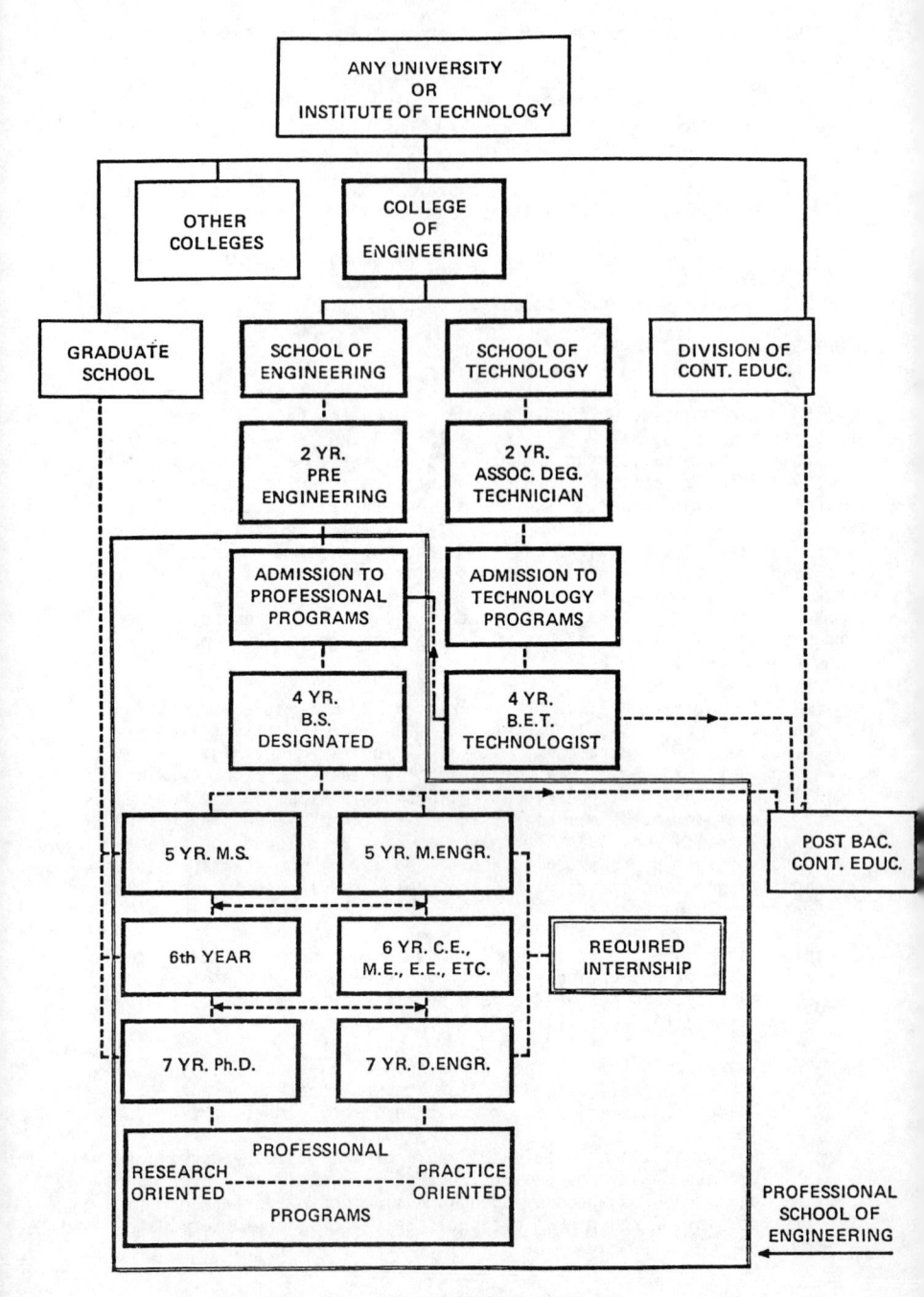

```
                    ANY UNIVERSITY
                          OR
                 INSTITUTE OF TECHNOLOGY

    OTHER                COLLEGE OF
   COLLEGES              ENGINEERING

GRADUATE      SCHOOL OF       SCHOOL OF       DIVISION OF
 SCHOOL      ENGINEERING      TECHNOLOGY      CONT. EDUC.

               2 YR.             2 YR.
                PRE            ASSOC. DEG.
            ENGINEERING         TECHNICIAN

            ADMISSION TO      ADMISSION TO
            PROFESSIONAL       TECHNOLOGY
              PROGRAMS          PROGRAMS

               4 YR.              4 YR.
                B.S.              B.E.T.
            DESIGNATED         TECHNOLOGIST

                                                POST BAC.
     5 YR. M.S.     5 YR. M.ENGR.               CONT. EDUC.

                    6 YR. C.E.,       REQUIRED
     6th YEAR      M.E., E.E., ETC.   INTERNSHIP

    7 YR. Ph.D.     7 YR. D.ENGR.

              PROFESSIONAL
    RESEARCH ---------- PRACTICE              PROFESSIONAL
    ORIENTED           ORIENTED                SCHOOL OF
              PROGRAMS                        ENGINEERING
```

ORGANIZATION OF PROFESSIONAL SCHOOL

the student's academic work. With appropriate qualifications and interest, the student could proceed into "advanced-level" professional school work along the "practice-oriented" path or the "research-oriented" path toward the M.Engr. or M.S. degrees, respectively. Again, depending on performance and interest, the student could terminate academic work or proceed toward the practice-oriented D.Engr. or research-oriented Ph.D. degrees.

4. On the "practice-oriented" path, an institution would have an option fo offering a six-year professional degree different from the M.Engr.

5. Corresponding points of progress on the two paths would involve approximately the same number of credit hours.

6. The D.Engr. degree would require an internship in place of the dissertation for the Ph.D. Course content would also reflect the differences in objectives.

7. As indicated in the diagram, graduate degrees also would be "housed" within the professional school but would be jointly administered by the professional school and the graduate school to satisfy the criteria.

COMMENTS ON MODEL

The model presented in Fig. 1 does not address all the questions stated earlier. It does, however, pose a minimum threat to existing programs and to the students, faculties, and institutions involved in engineering education. Indeed, one may ask if there is any significant benefit accomplished by the model. The following response is offered:

1. Pre-engineering qualifications are required before recongizing a student as formally admitted to the pursuit of an engineering degree.

2. A framework is established that leaves open the question of what degree represents the basic educational requirement for entry into the practice of engineering. However, it allows ready adaptation to future judgements of the profession.

3. The framework creates minimal disturbance to existing organizational structures. However, it would allow sufficient definition to permit an institution to have its school assessed for "recognition" or "accreditation."

It is not certain that the suggested model would satisfy all the NSPE recognition criteria.

No attempt is made here to consider specific institutional programs that may approach program structure and content that merit recognition as a professional school according to NSPE or other guidelines. It might be useful, however, to describe some of the actions taken at Texas A&M University. The doctor of engineering (D.E.) program* was established in 1974.[15,22,23] This program as presently structured includes the following elements:

1. Minimum Required Academic Work   (Semester Cr. Hrs.)
   Beyond B.S. Degree

   a. Professional Development   20 hrs.
   b. Seminar   4
   c. Internship (Report Required)   16
   d. Department-Oriented Graduate Level   32
   e. Engineering Design   12
   f. Elective Professional Development   12

                  Total Minimum Required   96

2. Undergraduate engineering students may enter the doctor of engineering program as juniors and continue work on their bachelor's degree.

3. A master of engineering examination is required at the completion of the student committee approved master's degree program.

4. A final comprehensive oral exam is required at the completion of the doctor of engineering program.

As of July, 1981, enrollment numbered 47 doctor of engineering students. The first D.E. graduated in December, 1975. Since that time, a total of 25 have received D.E. degrees. While it may be too early to predict the ultimate success of this program, it does appear that several tentative conclusions can be reached:

1. The program is attractive to students who wish to pursue advanced study that will enhance their practice of engineering as opposed to careers in research.

2. The intership thus far has proved to be not only practicable but also beneficial, and to have the interest and support from industry, government agencies, and engineering consultants.

3. Graduates of the program apparently are receiving salaries that reflect their employer's recognition of the D.E.'s special qualifications.

The doctor of engineering program requires an interview and satisfactory performance on the Nelson-Denny Test and the Minnesota Engineering Analogies Test before admitting students. Also, an Industrial

*The discussion that follows is taken substantially from Ref. 15, but with updated data.

Representative Committee composed of engineering practioners has played
a significant role in the development of this program.

While Texas A&M has not yet formally designated its engineering
programs to be a "professional school," various steps, including the
establishing of the doctor of engineering program already described,
have been taken in this direction. An assessment of the College of
Engineering in terms of criteria such as the NSPE recognition guidelines
has been undertaken. One of the products of this evaluation has been
a statement recently proposed by a college study committee concerning
criteria for "professional faculty." While no specific action has yet
been taken on this proposal, it is presented as Appendix C for possible
interest. It is perhaps noteworthy that the committee concluded from
an examination of backgrounds of present members of the college faculty
that a substantial majority of the faculty meet these criteria.

CONCLUSIONS

The purpose of the previous discussion is to offer some illumin-
ation concerning the concept of professional school of engineering, to
note some of the many questions that frequently arise in discussions of
the feasibility and desirability of implementing the concept, and to
comment on a possible model that has some attractive features.

For constructive actions to emerge, there needs to be unemotional
and objective evaluation of alternatives that would reflect the
collective wisdom of members of the engineering profession, including
engineering educators. Too often discussions have been clouded--and
hampered--by biases and a reluctance to pursue actively such an objective
analysis.

REFERENCES

1. Lohman, M.R., "Professional School for Engineers," *Engineering Education*, Jan., 1970, pp. 954-955.

2. McCollom, Kenneth A., "Professional Schools of Engineering," *Engineering Education*, May, 1972, pp. 915-918, 942.

3. Minutes of NSPE Annual Meeting, July, 1972.

4. Lohman, M.R., "President's Report to the Council," *40th Annual Report, Year Ending Sept. 30, 1972*, Engineer's Council for Professional Development, pp. 4-7.

5. Canjar, Lawrence N., "Examples of Professional School of Engineering," *Engineering Education*, Feb., 1972, p. 442.

6. Schaub, James H., "Professional Schools for Civil Engineering Education," *Civil Engineering Education*, ASCE Conference on Civil Engineering Education, Feb. 28-Mar. 2 1974, pp. 735-751.

7. McLellon, Waldron Murrill, and Robert Donovan Kersten, "The Professional Faculty," *Civil Engineering Education*, ASCE Conference on Civil Engineering Education, Feb. 28-Mar. 2, 1974 pp. 800-807.

8. Grinter, L.E., "Defining a Professional School of Engineering," (Guest Editorial), *Engineering Education*, Jan., 1975, pp. 279, 354-355.

9. Dillard, Joseph K., "Professional Schools of Engineering - A Call for Action," *IEEE Transactions on Education*, Nov., 1976, pp. 127-132.

10. Drucker, Daniel C., "On the Most Suitable Professional Schools of Engineering," *Proceedings, Seventh Annual Frontiers in Education Conference*, ASEE/IEEE, Oct. 24-26, 1977.

11. Pletta, Dan H., "The Development of Professional Schools," *NSPE Workshop on Professional Schools of Engineering*, Chicago, Illinois, Mar. 13-14, 1978.

12. Hall, Carl W., "Professional Schools of Engineering - Guidelines for Professional Schools," *1978 ASEE Annual Proceedings*, pp 168-170.

13. Kimel, William R., "The Feasibility of Professional Schools of Engineering in Missouri," *1978 ASEE Annual Conference Proceedings*, pp. 212-215.

14. McCollum, Kenneth A., "The Professional School Model at Oklahoma State University," *1978 ASEE Annual Conference Proceedings*, pp. 252-259.

15. Samson, Charles H., and Dan H. Pletta, "New Educational Environment for Tomorrow's Professional Engineer," *Proceedings, Eighth Annual Frontiers in Education Conference*, ASEE/IEEE, Oct. 23-25, 1978, pp. 208-215.

16. Rodenberger, Charles A., "The School of Professional Engineering: an Administrative Model," *Engineering Education*, Apr., 1981, pp. 671-676.

17. Matheny, Charles W. Jr., "Needed: A State Professional School of Civil Engineering," *Engineering Education*, Apr., 1981, pp. 684-686.

18. *Challenge for the Future...Professional Schools of Engineering*, prepared by the Professional Schools Task Force, National Society of Professional Engineers, 1976.

19. *Proceedings of the Workshop on Developing Guidelines for Recognition of Professional Schools of Engineering*, sponsored by the Task Force on Professional Schools of Engineering, National Society of Professional Engineers, Chicago, Illinois, Mar. 13-14, 1978.

20. Minutes of NSPE Winter Meeting, Jan., 1980.

21. "Report of Committee to Examine Long Range Objectives for Engineering Education," Professional Engineers in Education Division, NSPE, Jan., 1978.

22. Benson, Fred J., and Charles H. Samson, "A Proposed Doctor of Engineering Program," *Civil Engineering Education*, ASCE Conference on Civil Engineering Education, Feb. 28-Mar. 2, 1974, pp. 772-778.

23. Samson, Charles H., and Donald McDonald, "Experience With a Professional Doctorate," *Proceedings of ASCE Conference on Civil Engineering Education*, Apr. 19-21, 1979, pp. 323-329.

APPENDIX A

PP No. 104 -A -- Professional Schools of Engineering

It is the policy of the National Society of Professional Engineers to urge all segments of the engineering profession to seek the highest standards of preparation for engineering practice. The enhancement of engineering education is clearly in the interest of protecting the health, safety, and welfare of the public and therefore a worthy objective of a learned profession.

NSPE recognizes the need for broadened educational preparation of persons entering the engineering profession and the importance of innovative and creative approaches to the use of technology for the benefit of mankind within certain environmental constraints and yet provide improved quality of life for all. Recognizing the diversity of engineering education programs existing in the United States, NSPE does hereby adopt the following definition of a Professional School of Engineering:

> A Professional School of Engineering is a recognized educational unit which provides the formal education, beyond the baccalaureate degree, necessary to enter the practice of engineering.
>
> The Unit operates programs under the direction of qualified practitioners, with appropriate academic and non-academic experience, and provides elements of general, scientific and professional education within the guidelines established by the profession.

NSPE urges increased commitment of the profession to further the development of engineering educational units, including faculty, facilities, and programs, as they strive to meet the needs of society for well-qualified engineering manpower. Further, it is the policy of NSPE to grant formal recognition to Professional Schools of Engineering under guidelines adopted by the Board of Directors.

APPENDIX B
NSPE Guidelines for Recognition
of
Professional Schools of Engineering
(Adopted January, 1980)

INTRODUCTION

The stated purpose of the National Society of Professional Engineers as delineated in the Constitution, includes the following concepts related to education:

1. Service to Society, to State, and to the Profession is the premise upon which individual opportunity must be built.

2. Dedicates itself as an educational institution to the promotion and protection of the profession of engineering as a social and an economic influence vital to the affairs of men and of the United States.

3. Stimulates and develops professional concepts among all engineers.

4. Acknowledges individual achievement in engineering as relected by education and practice.

In seeking to accomplish these purposes, the Board of Directors has authorized the establishment of a program of "recognition" of Professional Schools of Engineering. Such recognition is to be granted by the Board of Directors to those engineering schools which, in addition to meeting appropriate accreditation criteria, have gone "the second mile" and further advanced the preparation of its students for professional practice in accordance with guidelines adopted by the Board of Directors.

DEFINITION

A Professional School of Engineering is a recognized educational unit which provides the formal education, beyond the baccalaureate degree, necessary to enter the practice of engineering.

The Unit operates programs under the direction of qualified practioners, with appropriate academic and non-academic experience, and provides elements of general, scientific and professional education within guidelines established by the profession.

Programs

1. Recognition shall be based on engineering education programs with a minimum duration of five years and which are *either* accredited at the advanced level by the Accreditation Board for Engineering and Technology (ABET) *or* accredited at the basic level by ABET and are capable of meeting advanced level criteria.

2. In addition to satisfying ABET quantitative criteria at the advanced level, NSPE recognition will expect the program to include:

   a. The equivalent of one-third year of study in a coordinated/planning area of instruction covering non-technical subject matter pertinent to the preparation for professional practice, e.g., engineering ethics, engineering law, communications, economics, accounting, finance, etc.

   b. The equivalent of one-third year of study in a coordinated/planned area of instruction in such a manner as to develop in all students management and leadership skills and a sensitivity to and understanding of the engineer's responsibilities for the public welfare and concern for natural resources.

   c. To provide supervised clinical experience during the formal educational process of at least six month's duration, such as internships, cooperative programs, graduate assistantships, or specialized fellowships involving work experience related to the major field of study and accomplished after admission to the professional phase of the program.

Admissions

3. Monitor university or college admissions programs to assure that individuals with appropriate preparation and abilities enter the basic phase of the program as exemplified by admission criteria, entrance examinations or standardized tests, and completion of appropriate courses essential to engineering study.

4. Admit students to the professional phase of the program after they have completed suitable college work in basic sciences, mathematics and engineering sciences. The academic work presented and the results of a personal interview must demonstrate a suitable level of competence so as to ensure successful completion of the professional phase of the program.

Students

5. Encourage students to participate and provide evidence that the students do participate in professional and extracurricular activities, university, and community affairs.

6. Encourage students to develop and maintain a concern for safety, welfare and health of the public and provide evidence that students are involved in formal seminars, field trips, case studies and appropriate clinical simulations related to the major field of study.

7. Specify that all professional degree candidates pass a terminal professional competency examination.

Faculty

8. Maintain a continuing record of contribution to the affairs and activities of an appropriate engineering society, and/or involvement in public and legislative affairs.

9. Provide evidence of a significant involvement of a majority of the faculty in non-academic engineering consulting and other up-to-date engineering experience gained through continuing practice, research, professional and sabbatical leaves, and other work experiences.

10. Require that faculty members having responsibility for teaching engineering synthesis and design courses be registered engineers in the state in which the educational institution is located.

Schools

11. Provide evidence of commitment of members of the profession from off-campus in the educational process including establishing educational objectives (e.g., adjunct faculty, seminar speakers, internship proctors, boards of visitors, advisory committees, etc.).

12. Provide programs for the maintenance of professional competence for practitioners.

13. Make provisions to ensure that student-faculty ratios and class size are appropriate to the instructional mode and provide evidence that the higher administration recognizes social needs through differential funding allocations.

14. Assure that appointment, evaluation and change of status (tenure, promotion, and salary) of engineering faculty shall be the responsibility of the administration of the engineering unit within the governance framework of the institution.

APPENDIX C

PROPOSED CRITERIA FOR PROFESSIONAL FACULTY, COLLEGE OF ENGINEERING TEXAS A&M UNIVERSITY

The Professional Faculty of the College of Engineering is composed of members of the faculty whose education and experience qualify them to plan and conduct programs which will prepare students for the professional practice of engineering.

Nominations for membership on the Professional Faculty shall be made by the Head of the appropriate department to the Dean of Engineering. It is recommended that a review committee be established to evaluate these nominations and to make appropriate recommendations to the Dean. Appointments to the Professional Faculty of the College of Engineering shall be made by the Dean of Engineering.

The Professional Faculty shall be composed of Members and Adjunct Members. Qualifications for appointment as a Member of the Professional Faculty are:

1. A Tenure-Track position in an academic department of the College of Engineering.

2. Nomination by the Head of the individual's academic department.

3. Registration as a Professional Engineer in the State of Texas.

4. Experience with the professional practice of engineering beyond a purely academic context. This experience may be through a full-time professional practice, consulting or sustained personal contact and professional interaction with people or agencies engaged in the professional practice of engineering. Professional practice is here understood to mean carrying a significant level of responsibility for a substantial engineering activity. It is characterized by education beyond the baccalaureate level, high technical and ethical standards, and a broad perspective in defining the parameters of an engineering problem and evaluating the efficacy of alternative solutions.

Adjunct members may be appointed who are not members of the faculty of the College of Engineering but are registered Professional Engineers and meet the experience qualifications listed above. Adjunct members may also be appointed who are faculty members but hold registration in another state or who are in a non-tenure track position but are eminently qualifieid by past experience. They are nominated on the basis of contributions they would be expected to make within the professional programs of the College and their professional qualifications with respect to these expected contributions.

Membership on the Professional Faculty shall be for a period of not more than five years. Reappointment will be based on continued to meet the qualifications for Membership and on participation in the professional programs of the College. Appointments to the Professional Faculty may be terminated at any time by the Dean of Engineering.

THE NEED FOR CONTINUING EDUCATION AND ITS QUALITY CONTROL

Session Leader:

Roy H. Mattson
University of Arizona

OUTLINE

I Background

    A The need for continuing education for engineers is obvious based on the money being spent on it by practitioners, industry, and government.

    B The need for quality control in continuing education is not documented.

    C There is agreement that ABET has provided useful quality control for undergraduate engineering education.

II Who benefits from continuing education for engineering practitioners?

    A The participant

        1. to avoid obsolescence
        2. to change field
        3. to meet licensing requirements
        4. for career development

    B The industry

        1. to develop new competences
        2. to improve engineering capabilities
        3. to compete

    C The public

        1. improved products
        2. improved services

III What is happening in other professions and countries?

    A Mandatory continuing education and relicensing

    B Mandatory continuing education for engineers in Germany and France.

IV Engineering Manpower Problems

   A Use continuing education to help meet them.

   B The technical societies should lead the way.

V   IEEE Validation Program

   A Motivate practitioners.

   B Monitor quality.

   C Maintain records.

   D Accept courses from new course sponsor.

   E Recognize evaluated courses with continuing education achievement units (CEAUs).

   F Recognize other courses with appropriate units (CEU, PDU, PDH).

   G Non-degree seeking persons taking regular courses from ABET departments can obtain CEAUs.

VI ABET accreditation of undergraduate curricula is useful.

   A to students.

   B to departments.

   C to others.

VII ABET accreditation of continuing education course sponsors is needed.

EFFECT OF COMPUTER GRAPHICS ON BOTH INDUSTRY AND ENGINEERING EDUCATION

by

Edward M. Rosen
Fellow
Monsanto Company

Summary

The explosion in the use of computer graphics in industry is being fueled by the promise of substantial productivity gains. These gains in turn are being made possible due to the decreasing cost of hardware, the availability of suitable software, the advent of microprecessing technology and despite the ever rising costs of manpower.

The recognition that graphics provides a natural means of communicating with the machine has opened the door to wider uses of the computer in engineering design, science and business. Integrated systems in Computer Aided Engineering (CAE) imply the need for broadly trained engineers who are familiar with a range of engineering disciplines which in turn increase the importance of the service courses taken outside the engineering major.

A major challenge to the engineering educator is the design of courses to properly prepare the student to use the natural language of graphics in engineering analysis, business, science and in the increasingly integrated engineering system being used and developed in industry today.

1. The Revolution in Computer Graphics

The beginnings of modern computer graphics is considered to have had its origins in the SKETCHPAD project at MIT in 1963 (1,5). Despite the pioneering work done in the automotive and aircraft industries during the 60's and early 70's a Battelle Computer Graphics Conference entitled "Why is Computer Graphics Always A Year Away?" was held in Seattle July 31-August 2, 1973. A number of things have happended since that time, however.

1. The cost of hardware has dropped dramatically.

2. Quality graphics and data base software has become available.

3. On line systems have become more cost effective.

4. The personal computer has arrived.

The result, especially since about 1980 has been an explosion in computer graphics. Articles attesting to this interest have appeared in BUSINESS WEEK (6), THE HARVARD BUSINESS REVIEW (7), as well as the WALL STREET JOURNAL and numerous technical publications. The future of computer graphics in fact appears only to be limited by our creativity.

Figure 1 gives the current and expected revenues in the computer graphics market place. The 25 percent growth rate is expected to result in a 4 billion dollar market by 1985 (8).

Hardware and software technologies in graphics will continue to improve by taking advantage of the large base of TV technology (4) that has developed both in the receiver area as well as in TV recorders. Cable TV will no doubt play an increasingly important role in distributing graphics broadly.

2. The Graphics Milieu

The greatly expanded interest in computer graphics has given rise to an environment consistent with other high technology fields. Figure 2 gives some of the current technical groups and publications that are concerned with computer graphics. Large numbers of seminars (University sponsored as well as by professional groups) summer courses, shows and conventions are held throughout the year. Enterpreneurial hardware, software and consulting businesses have proliferated.

3. Graphics as a Natural Language

In much the same way that mathematics is a natural language of engineering so too is graphics through the medium of sketches, drawings, pictures, plots and blueprints. These graphical methods pervade the entire field of engineering and business. But while digital computation provided the means of communication via mathematics the lack of suitable hardware and software in the past has limited the full utilization of graphics as a means of communication.

Graphics as a means of communication between man and machine has extraordinary power (witness the impact of TV) and does then provide a definitive opportunity for productivity.

The use of graphics to gain an understanding of a phenomena is common. But the sheer magnitude of plotting many points often may be discouraging. Figure 3 is a plot of the traction (10) in a fluid which gives an understanding which a table of numbers simply could not. Figure 4 (11) also gives insight which would be difficult to achieve in any other way.

As with any means of communication graphics can be used for the specific purpose of the presenter (12). For example, The Johnson & O'Rouke Manufacturing Company has amassed statistics on the sales volume of its chief product, the Clasmotron, which, incidentally is the object of a government antitrust suit. Let us see how various groups might report this information.

The Computer Department, bogged down with work, wants to minimize its effort, so it simply produces computer listings. This gives the sales in dollars for each quarter since 1969 of Johnson & O'Rouke and of its competitors. The listings quickly find their way to the filing cabinet.

The Sales Department (Figure 5) wants to show how well it has done. It decides to plot a trend line (linear regression) of the percent of sales (market coverage) over the recent years. The trend line, pointing staunchly upwards, creates expectations of continued growth.

The prosecuter in the antitrust case (Figure 6) uses the same information but he selects only the final years data and a pie chart to present the data. The heavy shading, contrasting with the omitted pie segment indicating the remaining market share makes Johnson & O'Rouke appear predatory.

The defense attorney (Figure 7) wants to emphasize the significant market share covered by the competition. He therefore shows sales in dollars over the entire period from 1969 to the present, making the competition look stronger than it is.

The president of Johnson & O'Rouke reports to the stockholders that the company has performed very well, "much better than the competition" in the last few years. To back this contention he prefers to show percentages. Johnson & O'Rouke (Figure 8) is shown as a solid curve, suggesting much more substance than the puny (dotted no less) curves indicating the competition. He also includes data from the dark ages (back to 1969) to emphasize the improvement made in recent years (since he took over).

The management consultants brought in by the president who provately fears things may not be as good as he claims, presents the situation as it is: Johnson & O'Rouke is dominating a dwindling market. The Clasmotron is obsolete: the company must concentrate on inventing a new product, not on improving the Clasmotron to beat the competition (Figure 9).

Finally, an antiregulation lobby gets hold of these data. They don't care about Johnson & O'Rouke, they just want to emphasize the fact the president chose to ignore: the death of the Clasmotron (Figure 10).

Since graphical representation is a natural language and since the hardware and software is now available for its use, the need to use the language well will become more important.

4. Computer Graphics in Engineering

Computer graphics systems generally can be thought of as having three major components (7).

1. Hardware devices

2. Software

3. Data Bases

Depending on the application and the role of graphics in it, each of these components may be of varying importance. The application also determines how the user interacts with the system and therefore the nature of the required hardware devices.

The area of CAD/CAM (Computer Aided Design/Computer Aided Manufacturing) is generally concerned with the design of a product. Here the need is for highly interactive graphics, the need to modify drawings and sketches and the need for high quality displays. Physical shape is of central importance, drawings are the means of communication and analysis can be generalized. Data bases may or may not be of great importance depending on how integrated the system is with other applications.

In chemical engineering (9) the area of CAPD/CAPM (Computer Aided Process Design/Computer Aided Process Management) is concerned with the design of the process to make the product. Computer based models are used to aid in the design of the process and its operation. The structure of the process model is of central importance. Process flow heat and material balances are the means of communication. Graphics needs are more passive. Analysis are more process specific. Process models are used for strategic planning, process development, process design and process operation. Data base needs are often specific (physical properties).

Systems which attempt to intergrate engineering design workflow (process document retrieval, design document retrieval, construction documents), project accounting (project costs, invoice verification) cost engineering and project scheduling are often referred to as CAE

(Computer Aided Engineering) systems. Here data bases are of great importance, passive graphics is needed (as for generating critical path reports) and adequate software currently rarely exists. The potential for productivity gains here is significant and the area is one of current interest to the chemical industry.

Drafting systems are often implemented as a forerunner of much larger CAD/CAM or CAE systems. Here hardware is often minimal, graphics is of central importance and data bases are needed to store and retrieve drawings.

The area which has become known as business graphics is one of the fastest growing today (52% annual growth rate projected through 1985). The spurt in this area is largely due to the availability of a high level language so the manager does not have to depend on a computer programmer. Here the need is for business data bases which interface with graphics system. The need is for graphical output systems as opposed to graphical input systems.

5. The Promise of Productivity

To date, office productivity improvements have focused largely on improving the output of clerical and support staffs. Graphics directly attacks the productivity of the knowledge worker whose employment cost is several times that of the clerical and support staff.

Graphics, used properly, does not cause productivity, it allows productivity. The rising popularity of computer graphics in industry attests to the fact that it can be used properly and there are real gains in productivity. This, despite the fact that productivity has many faces and is often difficult to assess and measure.

Experience at the Monsanto Company has shown that with a computer drafting system (to do drawings only) two man hours of drafting can be replaced with one hour of machine time (one operator). Without considerable previous efforts on standardization this would have probably resulted in a ratio of three to one, an often states industrial norm.

Engineers and scientists using online graphical systems for engineering and data analysis report productivity gains ranging as high as 8 to 1 when measured in terms of elapsed time. Cost/benefit studies uniformly point to the advantages of online computation coupled with graphical output. The productivity aspects of better design and analysis due to graphics, though often intangible, are nevertheless real.

6. Computer Graphics in Engineering Education

Though University research has contributed to the development of computer graphics, the integration of the use of graphics into the engineering curriculum has been slow. This has been mainly due to the cost of the graphics hardware (3). Other problems are the availability of instructors suitably trained and interested in the topic. The need is for an instructor who is willing to implement, support and document suitable software to carry out his desired educational goals. Adequate support is often not available through the University computer center.

There has been some activity, however, which has attempted to address the problem of the role of computer graphics in engineering education. The CACHE Corporation (Computer Aids in Chemical Engineering) task force on graphics has published a report in 1978 (2) to help faculty members become interested and started in the field. A recent survey article (3) reported on the graphics activities in a number of chemical engineering departments around the country. Recently the National Computer Graphics Association reported on efforts to work with a Consortium of Universities to explore the general problem (re: Dean Gibson, Univsersity of Virginia).

Currently Professors Edgar (Texas), Reklaitis (Purdue) and Mah (Northwestern) are working on an ASEE Commission Study (re: Dean Glover, Ohio State) on the Impact of Advanced Techonology on Engineering Education which will address computer graphics in chemical engineering education.

7. Summary and Recommendations

Computer graphics is a natural language for the student and practicing engineer in industry to communicate with the machine. As a result his potential for productivity is substantial. In order to prepare the student for use of this modern technology several things can be done:

1. Encourage the student to gain typing skills to use a keyboard.

2. Encourage the student to view data graphically, when at all possible (by hand if not by machine).

Educational institutions might consider some of the following thoughts:

1. Faculty members should be encouraged to acquaint themselves with this technology by attending shows, acquainting themselves with vendors and finding our how it is being used. Participation in CACHE (or other organizational) task forces is suggested.

2. Funding requirements can possibly be approached by participating in consortiums. This can lead to sharing of software as well as approaching software houses to make their software available for educational uses. (Possible: Approach ISSCO for their TELLAGRAF and DISSPLA packages. Lower level packages are available from CALCOMP and TEKTRONIX).

3. Personal computers have very useful graphics capabilities. A good start in graphics at low cost can be made using these machines.

Literature

1. Potts, J., "Computer Graphics - Whence and Hence", COMPUTER & GRAPHICS vol 1, no. 2/3, (Sept. 1975), p 137-156.

2. Carnaham, B., Mah, R.S.H. and Fogler, S., "Computer Graphics in Chemical Engineering Education", CACHE CORPORATION, University of Utah.

3. Edgar, T.F., "Computer Graphics in Chemical Engineering Education" CHEMICAL ENGINEERING PROGRESS, vol 77, no. 3 (1981), p 55-59.

4. Brown, B.E. and Levine, S.L., "The Future of Computer Graphics", BYTE Publications Inc. (Nov. 1980) p 22-28.

5. Benest, I.D., "A Review of Computer Graphics Publications", COMPUTER & GRAPHICS, vol 4 (1979) p 95-136.

6. Business Week, "The Spurt in Computer Graphics", June 16, 1980.

7. Takeuchi, J. and Schmidt, A.H., "New Promise of Computer Graphics" HARVARD BUSINESS REVIEW, Jan-Feb 1980, p 122-131.

8. The Harvard Newsletter in Computer Graphics, vol 3 No 2, Jan. 26, 1981.

9. Evans, L.B., "The Future of Computer Aided Process Design" ASPEN USERS MEETING, May 1981.

10. Daniels, B.K., "Non-Newtonian Thermo-Viscoelestic EHD Traction from Combined Slip and Spin", ASKE TRANSACTIONS, vol 23, no 2 p 141-154.

11. Jolls, K.R., "Research an an Influence on Teaching", Department of Chemical Engineering, Iowa State University.

12. Rawlins, M., "Figures Lie & Liars Figure", ISSCO Seminar, Feb. 1981.

APPLICATION	REVENUE 1980	MILLIONS 1985
ELECTRONIC DESIGN	360	790
MECHANICAL DESIGN (CAD/CAM)	350	1600
BUSINESS	75	600
DRAFTING AND CARTOGRAPHY	200	290
CONTROL AND SCIENTIFIC	125	200
ANIMATION ART	28	45
EDUCATION, TRANSPORTATION, ETC.	280	540
	$1,418	$4,065

Figure 1 Revenues From Computer Graphics

Figure 1

GRAPHICS ORGANIZATIONS

ASSOCIATION FOR COMPUTING MACHINERY
    SPECIAL INTEREST GROUP ON GRAPHICS (SIGGRAPH)

NATIONAL COMPUTER GRAPHICS ASSOCIATION, INC.

WORLD COMPUTER GRAPHICS ASSOCIATION

INTERNATIONAL FEDERATION OF INFORMATION PROCESSING
    WORKING GROUP 5.2 ON COMPUTER AIDED DESIGN

HARVARD LABORATORY FOR COMPUTER GRAPHICS

GRAPHICS PUBLICATIONS

COMPUTERS & GRAPHICS (Pergamon Press)

COMPUTER GRAPHICS (SIGGRAPH)

COMPUTER AIDED DESIGN (IPC SCIENCE AND TECHNOLOGY PRESS Ltd.)

COMPUTER GRAPHICS AND IMAGE PROCESSING (ACADEMIC PRESS)

THE HARVARD NEWSLETTER ON COMPUTER GRAPHICS

SID JOURNAL (SOCIETY FOR INFORMATION DISPLAY)

Figure 2

TRANSVERSE TRACTION AT 1.00% SPIN

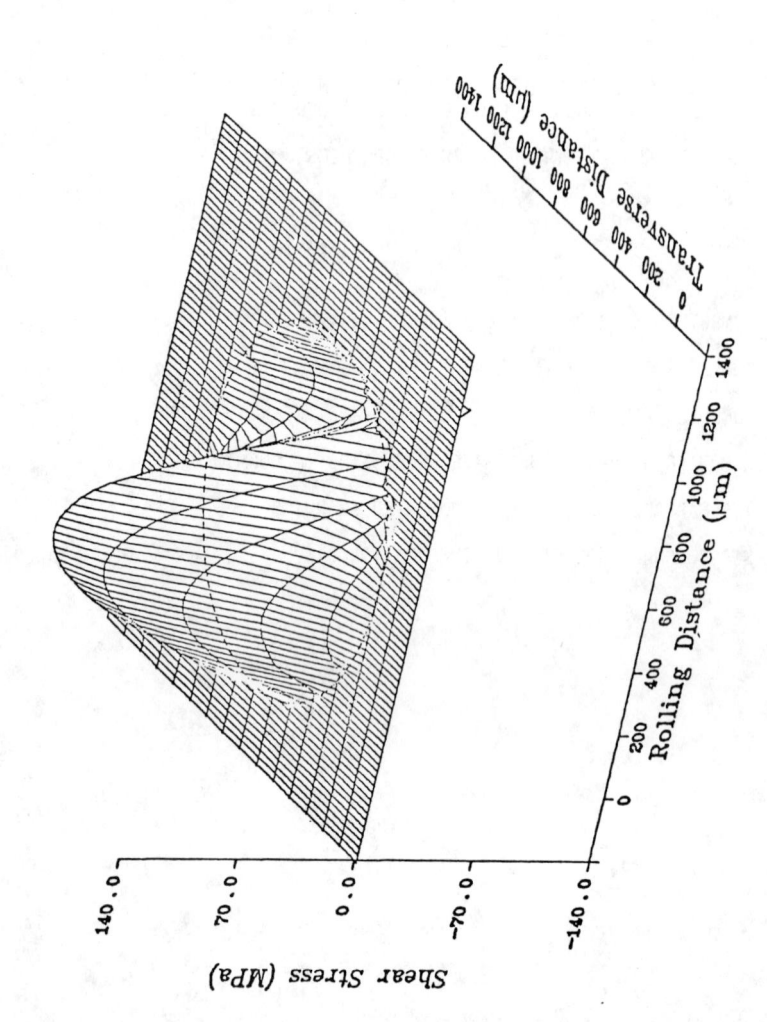

Figure 3

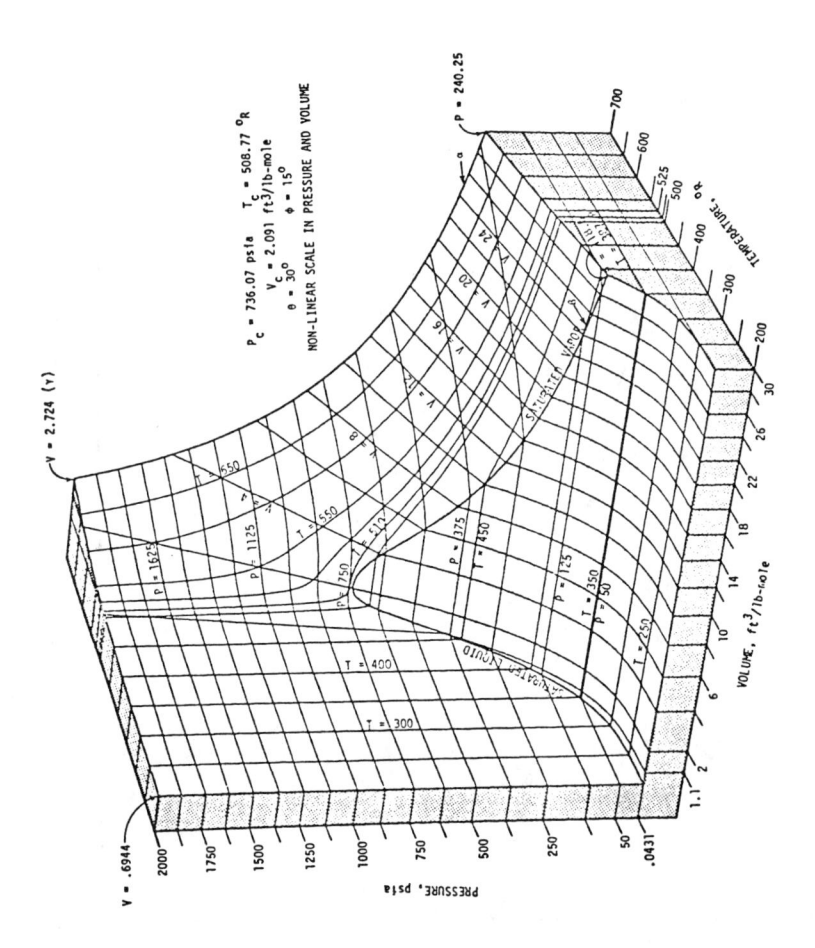

E. M. Rosen

Figure 4

Figure 5

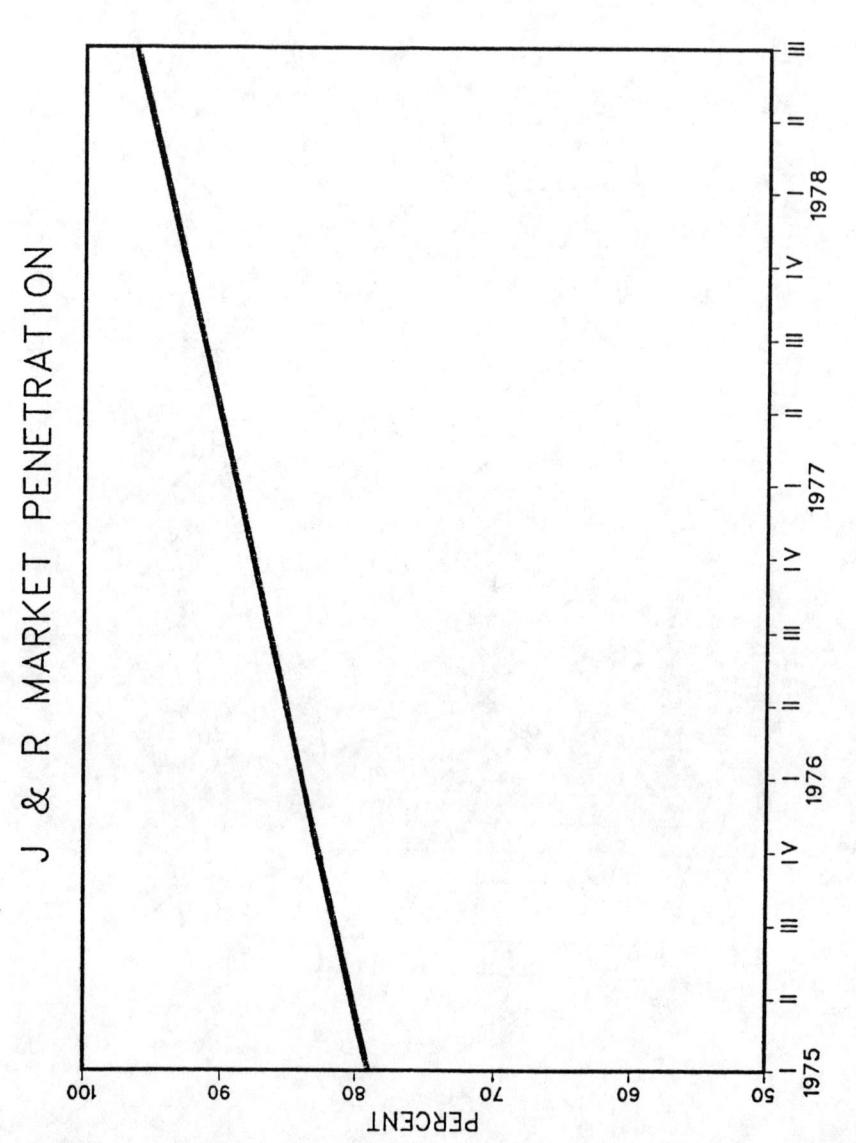

MARKET DOMINATION BY DEFENDANT

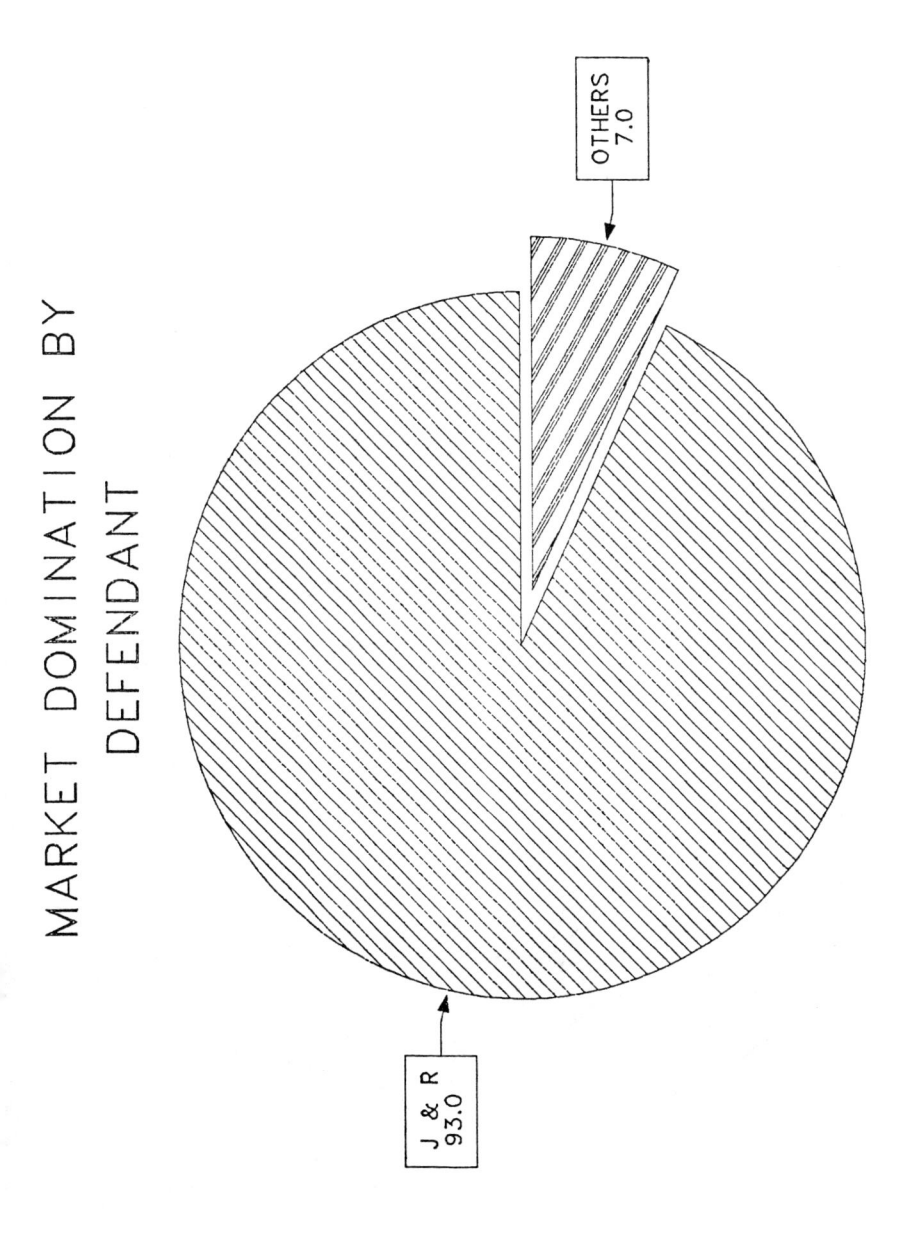

OTHERS
7.0

J & R
93.0

Figure 6

Figure 7

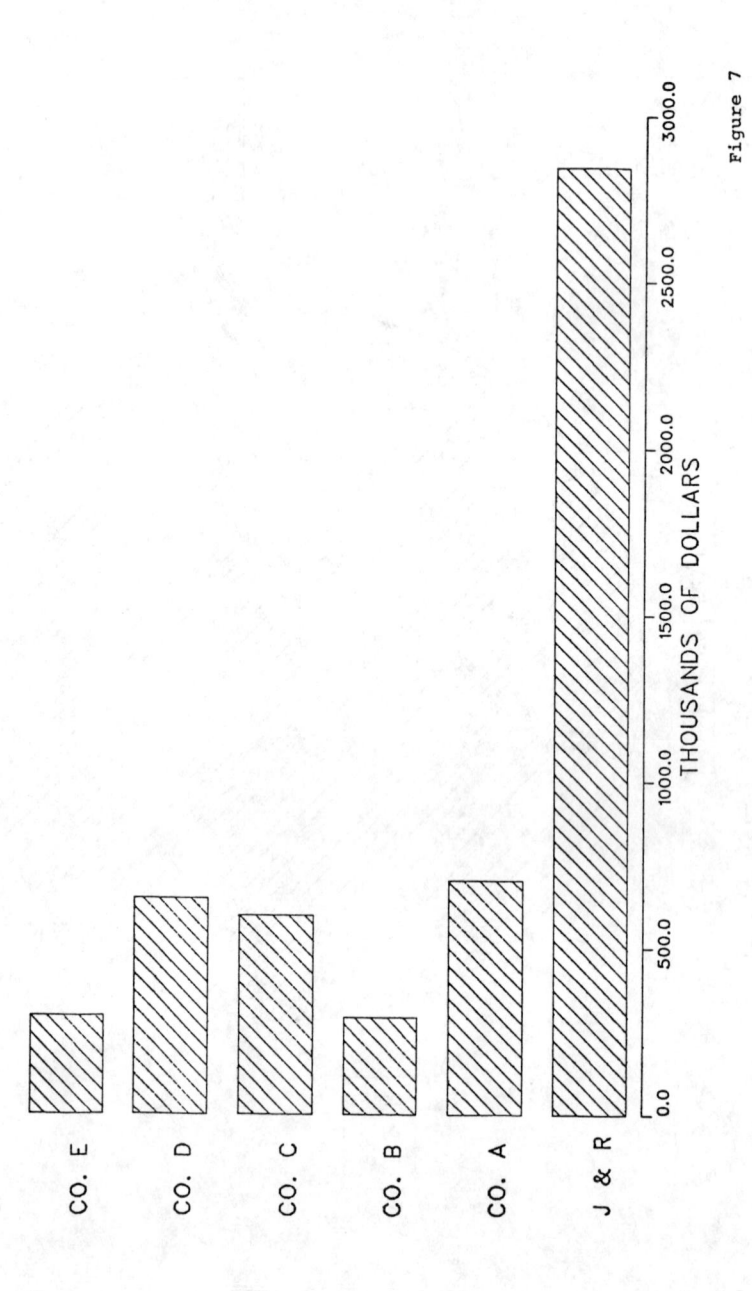

Figure 8

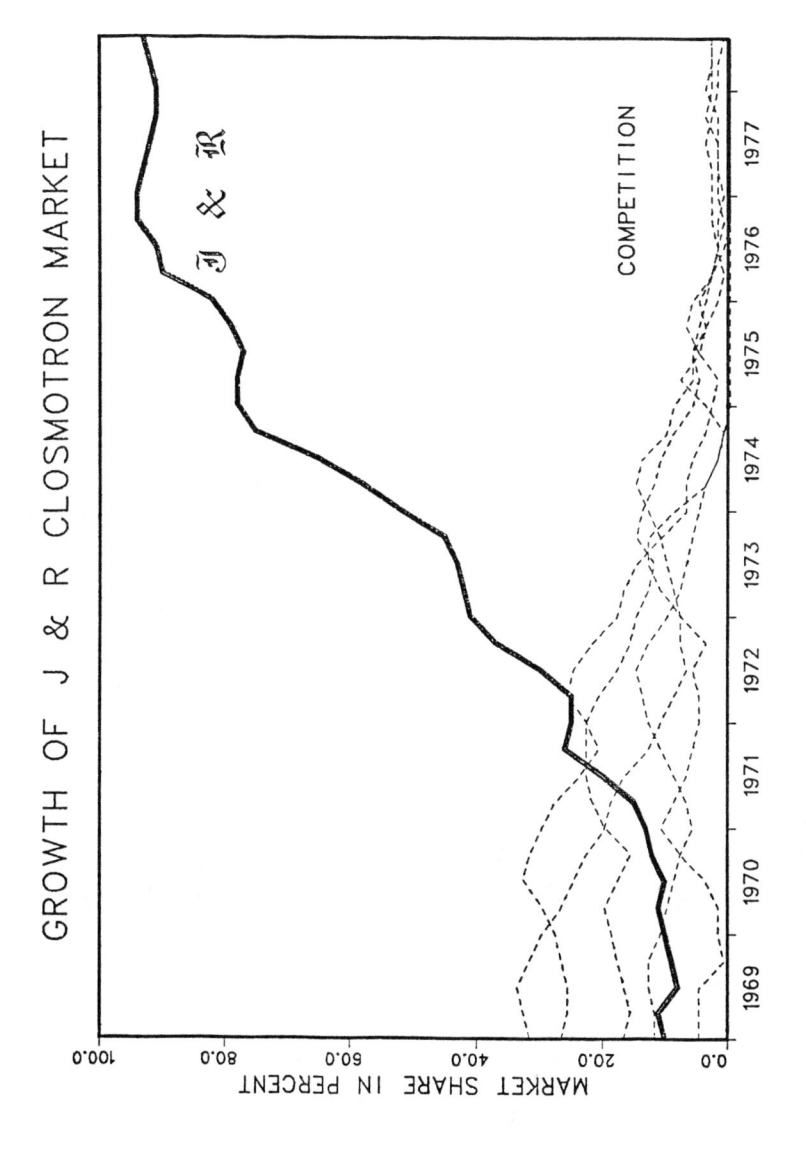

Figure 9. OBSOLESCENCE OF THE CLOSMOTRON

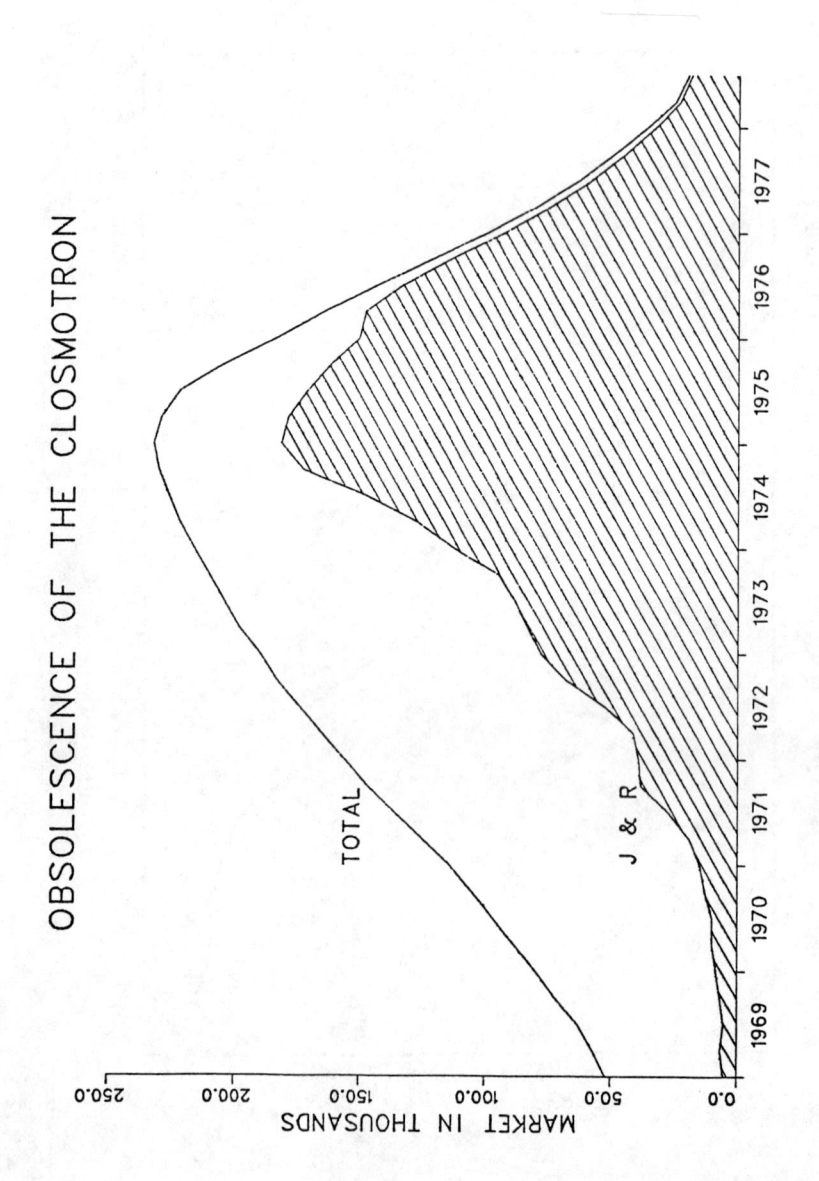

Figure 10

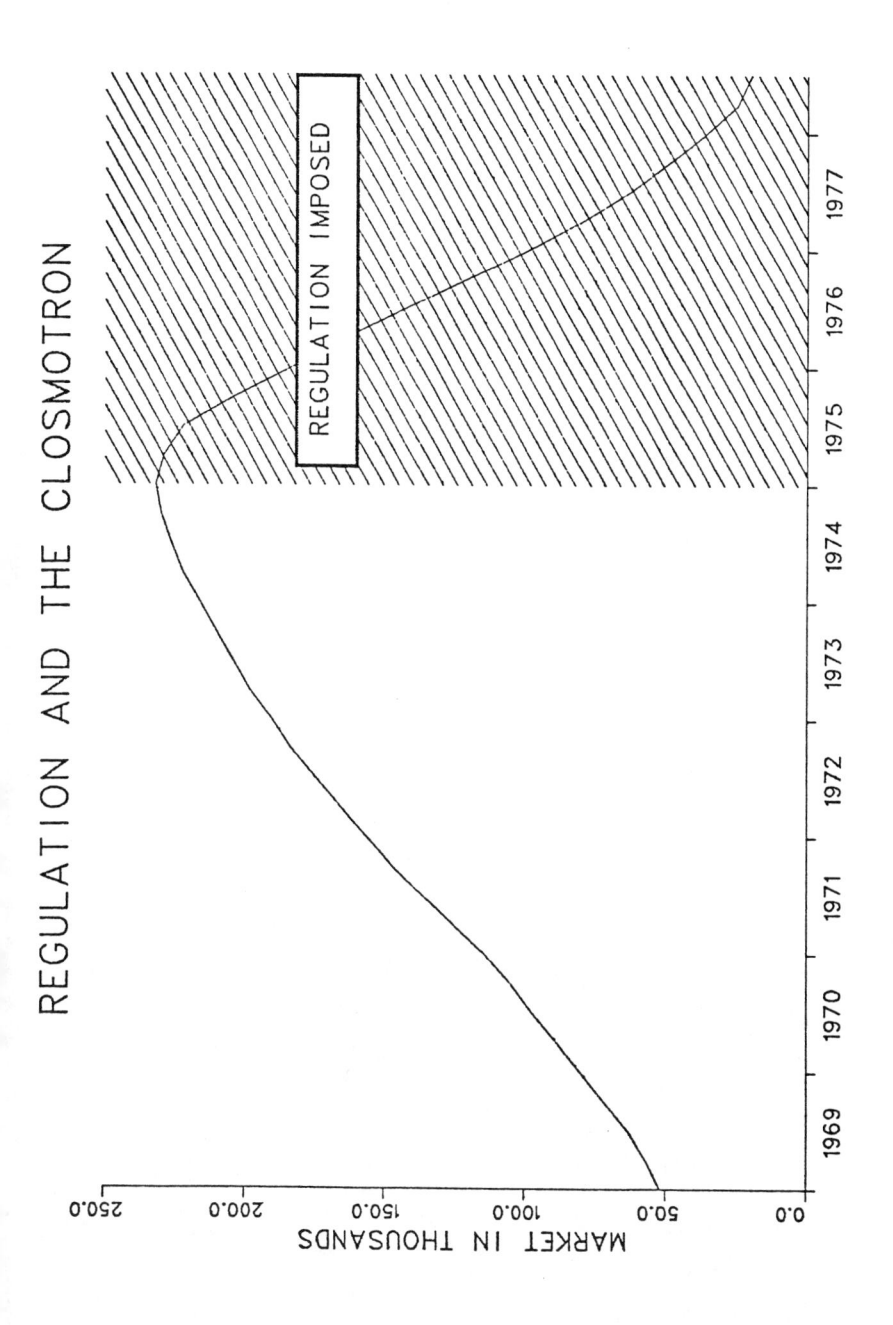

EDUCATION FOR ENGINEERING MANAGEMENT IN THE EIGHTIES

by

Merritt A. Williamson, P.E.
Orrin Henry Ingram Distinguished Professor
of Engineering Management, Emeritus and
Director, Engineering Management Program
Vanderbilt University

In the opinion of many of us who have held managerial positions in industry and government and who are now teaching engineering management, no discussion of Engineering Education - Aims and Goals for the Eighties would be complete without serious consideration of the emerging discipline of engineering management.

For far too long engineers have absorbed their managerial notions and attitudes informally from those with whom they have worked. It took many years before engineering was recognized as a legitimate university study. The university recognition of management is of much more recent origin. Both, however, are now recognized as comprising educationally communicable subject matter. Although management is essentially "doing" and not talking or writing about "doing", why should engineers continue to "muddle through" when they can learn many fundamentals which would enable them to perform their jobs with greater effectiveness.

The idea that all engineers who develop an interest in management should go to a business or management (B/M) school has been, and still is, quite prevalent, but it is a dated notion. It has been well demonstrated that this is not the most effective route for technically trained people to follow in order to improve proficiency in the management of technical people performing technical work in a predominantly technical environment. In 1966 there were only about ten graduate programs in the United States leading to the Master's degree in Engineering Management. There are now between ninety and a hundred such graduate programs and around thirty undergraduate programs, two of which are accredited by ABET. These programs have been developed within engineering schools in response to the demands created by the inadequacy of the business and management schools which have been with us since the turn of the century.

Many of us in the field are continually distressed by the lack of information in industry and government about programs in engineering management. There are also many universities that have little or no information on the existence of these programs. I, therefore, have taken this opportunity to discuss engineering education in the eighties.

Attached is a copy of the draft of an article which I was asked to prepare for the Third Edition of the Encyclopedia of Management, edited by Carl Heyel and published by Reinhold. This provides formal definitions, a brief history, and a discussion of the relationship to MBA programs. I shall not repeat this information here. Rather I shall address curriculum problems.

As a former Dean of Engineering (Penn State 1955-66), I can easily imagine the despair with which any engineering school will greet the thought of adding more requirements to an already crowded curriculum, whether undergraduate or graduate. Let me discuss this briefly.

There are two main questions as I see it. First, is there a need for adding studies in management to an already full curriculum in order to prepare students for the effective practice of engineering? Second, how can this need be met in the light of increasing demands for more technical knowledge in the various fields of engineering?

Many faculty members question the need for engineering students to know anything at all about management. They see management as being quite apart from the practice on engineering, and think of it as something which can be picked up later in some way not clearly defined. Because many faculty members have never held any managerial responsibility of any magnitude, this attitude is understandable. Those who have been successful managers in government and industry have arrived there without formal study and hence no demands on educators are made by them because they do not think in terms of collegiate preparation for this aspect of their practice.

A recent analysis of the work done by licensed engineers reports two primary job responsibilities: Adminstration/Management (reported by 34% of those surveyed) and Design (reported 24%). Design, as a primary responsibility, should come as no surprise, but when a higher percentage of licensed practicing engineers report that Administration/ Management takes precedence over design as their primary job responsibility, this should be given very serious consideration in any discussions of the Aims and Goals for Engineering Education in the Eighties. The data referred to comes from a study made on a sample of about 5000 licensed engineers and carried out by the National Council of Engineering Examiners. The report is entitled "NCEE Task Analysis of Licensed Engineers" and has just been released. It may be obtained by writing NCEE at P.O. Box 5000, Seneca, SC 29678. The price is $10.00.

Their findings agree with data presented in the September, 1973 issue (No. 25) of the Engineering Manpower Bulletin put out by the Engineering Manpower Commission of the Engineers Joint Council. This bulletin reported that an analysis of the 1969 National Engineers Register showed that 64% of the engineers responding reported that they had direct supervisory and managment responsibility. The bulletin also reported that in the 1972 postcensal study of the Engineering Profession conducted by the National Science Foundation, 480,000 people with college degrees in engineering classified themselves in other occupational categories and that managers and administrators made up a large part of this group.

The Engineering Manpower Commission Bulletin summarizes the issues very well:

"If management is an inherent part of the work of so many engineers, that fact should receive greater recognition in the education and career development of members of the profession. The typical undergraduate curriculum in engineering is almost entirely technical in nature, and most engineering students are oriented toward technical duties. Such a preparation is adequate for the first few years out of college, but not for the long run career requirements of a majority of engineers. Although more and more engineers are returning to school for advanced courses in business administration, only about two percent of all engineers held MBA degrees in 1969. The disparity between the formal education of engineers for management and the extent to which they actually become involved in it is apparent.

"Not only do engineers need more education in the arts and techniques of management, but they must be made more aware of they ways in which their duties will change during their professional careers. The gradual transition from technical to managerial responsibilities is something that more engineers should recognize as a normal pattern of career development. Too many engineers today are educationally and psychologically unprepared to make the shift, and seem to feel that they are abandoning their profession when they move out of strictly technical work. Young engineers should be helped to recognize that the current job structure, especially in industry, does not provide openings for all those who would like to stay in purely technical and nonsupervisory positions throughout their careers. Since the job pattern has developed over a long period of time in response to employer's needs, and is not traditional in many organizations, it is unlikely to be altered radically in the near future.

"It appears to be both normal and appropriate for a majority of engineers to encounter managerial responsibilities at some stage in their careers, and for the proportion of managerial duties to increase as the scope of the job increases. Not all engineers need to become managers and not all engineers will, but those who do are no less members of the engineering profession than those who do not."

Therefore, in answer to question one, the need seems to be well established. This now leads to the second question. How can this need be met?

There are ways which should be considered and discussed. First, courses in engineering management can be added to the engineering school offerings. A number of textbooks are available now to facilitate this. Such courses could be offered as either electives or requirements in each course of engineering study, but there are difficulties. One is reminded of the old quip that it is easier to move a cemetery than to change a curriculum! Another difficulty is to find truly competent persons to teach these courses. To teach management to engineering students requires more than placing a textbook in the hands of an engineering teacher regardless of how competent he may be in his field of engineering or how good a teacher he is. There is no substitute, in my opinion, for having had first-hand experience with actual managerial responsibility; and this management experience should be related to technical work. To the best of my knowledge only Industrial Engineering and Engineering Management Departments presently offer courses that would meet the requirements. Then comes the question, if this is so important to preparation for the practice of engineering, shouldn't such courses be required rather than be available as electives? Requiring a course or courses would compount the already existing shortage of qualified teachers. In some locations adjunct professors can be effectively used. It has been my experience that our present day students relate very well to teachers who bring into the classroom the flavor of the working environment. It may prove easier for adjunct professors to meet the needs of graduate students than to meet the needs of undergraduates, many of whom have not yet had any relevant experience.

Formal university courses, however, are not the only way to make managerial knowledge available. The Engineering Manpower Commission Bulletin, referred to above, suggests a continuing education approach. Some help can be obtained from short courses and seminars, as well as evening courses and in-plant educational programs. The efficacy of these options should be studied. Graduate courses which are especially designed for the practicing engineer have proved more effective than the general management courses which are offered in B/M schools. Witness the nine-fold growth of graduate programs over the past fifteen years! Most of these allow part-time study.

In summary, it seems obvious that any discussion of engineering education in the eighties would be incomplete without addressing this demonstrably important area. Those of us in Engineering Management education have been meeting the challenge, but the nature and the importance of our programs is not yet widely recognized. The Engineering Management Division of ASEE dates back to 1972, and the American Society for Engineering Management was founded in 1979 in response to a perceived need for professional affiliation of engineering managers regardless of their fields of technical expertise. The Professional Engineers also has a working committee on Engineering Management and Professional Leadership. When one also considers that many of the engineering societies comprising the American Association of Engineering Societies (AAES) have established management committees or divisions it seems obvious that education in engineering management deserves the timely and thoughtful consideration of leaders in engineering practice and in engineering education.

"From THE ENCYCLOPEDIA OF MANAGEMENT, 3rd edition, edited by Carl Heyel. Copyright 1982 by Van Nostrand Reinhold Company. Reproduced by permission of the publisher."

FOR ENCYCLOPEDIA OF MANAGEMENT, THIRD EDITION

ENGINEERING MANAGEMENT

ENGINEERING MANAGEMENT is the name given to the area of engineering and management concerned with managing technical work and technical people in a predominantly technical environment. This type of management is as old ad technology itself, but only recently has been recognized as an area deserving of a separate name and a separate educational preparation. The adjective "engineering" in this context is used to describe activities based on mathematics and the physical sciences, and is not narrowly restricted to the traditional engineering disciplines. Persons engaged in "Engineering Managment" may come from educational backgrounds other than engineering, but they must be qualified by education and/or experience to make sound decisions involving technical work.

FORMAL DEFINITION As with any area still seeking to define itself, no universally accepted definition has yet evolved. However, the most commonly accepted definition is:

> Engineering Management is the art and science of planning, organizing, allocating resources, directing and controlling activities which have a technological component.

This definition identifies a management specialty and also identifies specific activities which are integral to the full practice of the various engineering and scientific disciplines. The practitioners of engineering management, known as "Engineering Managers", generally claim an identification with some field of engineering or science or some related area which is also rooted in mathematics and the physical sciences. Engineering Management differs from Industrial Engineering to which it is most closely related on the engineering side, by its greater focus on "people" problems rather than on system design, which, of course, also includes people along with materials and equipment. (See Industrial Engineering). On the management side it differs from general management in its requirement that practitioners be competent in some technical field. Engineering managers may be found occupying top, middle, and supervisory management positions. (See Management Levels). They may be found working wherever a blending of managerial and technical knowledge is required, which may be in any organization, whether or not its primary business is technological.

HISTORY The management of technical activities by technically competent persons represents nothing new. It has been customary, however, for practitioners to identify themselves as either engineers or managers depending on their work environment. The establishment of Engineering Managment as a separate branch of academic study is a post World War II development. Graduate level courses in Research Administration were offered at Illinois Institute of Technology, New York University, and the University of Pennsylvania in the 1940's and early 1950's. Engineering management graduate programs leading to the master's degree were started during this same period at several institutions including George Washington University, New Jersey Institute of Technology, Rensselaer Polytechnic Institute, and the University of California in Los Angeles. It was also in the 1950's that the first engineering society, the Institute of Electrical and Electronics Engineers (IEEE) organized a unit to be concerned with engineering management. The American Society for Engineering Education in 1972 established its present Engineering Managment Division. The first real move toward recognition as a separate profession took place in 1979 when a group of interested people from government, industry, and universities organized the American Society for Engineering Management. The rapid growth of this society in its first year is indicative of the widespread interest in this field.

EDUCATION Although education alone does not qualify one for the responsible practice of management or engineering, there has been a growing recognition that engineers need a more extensive knowledge of management. Survey data from the Engineering Manpower Commission collected in 1969 showed that more than 80% of all the engineers surveyed were regularly assigned managerial duties. (Engineering Manpower Bulletin No. 25, September, 1973). The Bulletin stated, "If management is an inherent part of the work of so many engineers, that fact should receive greater recognition in the education and career development of members of the progession....Not only do engineers need more education in the arts and techniques of management, but they must be made more aware of the ways in which their duties will change during their professional careers." Universities and colleges have responded by offering classes and establishing programs leading to degrees. In addition, short courses, seminars, and workshops are being offered in increasing numbers. As of 1980, students in over 30 institutions may enroll as a candidate for the bachelor's degree in Engineering Management studies. (Most of these curriculums are known by other names such as Engineering Administration, Engineering Operations, Management Engineering, etc.) This diversity of name, which is also found, but to a lesser extent, in programs at the postbaccalaureate level, is a reflection of the recent application of a suitable name to distinguish this important area of professional practice. As of 1980, two of the baccalaureate curriculums have been accredited by the Accreditation Board for Engineering Technology (ABET). Undergraduates receive an education in engineering fundamentals and in management-related subjects. A more extensive response to needs has been the establishment of master's and doctor's programs. In 1980, these programs numbered over 80. Many of these are offered on a part-time basis. Most of them prefer and may require applicable work experience prior to entry.

RELATIONSHIP TO MBA PROGRAMS Business and Management Schools offer programs leading to the Master of Business Administration (MBA) degree. These programs usually require two academic years to complete and impose no requirement of prior work experience. Some of them provide for specialization in such areas as accounting, banking, marketing, etc. Engineering Management graduate programs usually require one year of formal study following a technical undergraduate education. This year of academic study may be taken on a full or part-time basis. The curriculum is usually a Graduate School, and is designed to provide additional studies of particular importance to those who aspire to managerial responsibility where technical knowledge is also required. The MBA and Engineering Management programs are in no way competitive, because they are designed to meet entirely different needs.

ACCEPTANCE AND POTENTIAL Recognition of this **blend of engineering** and management as a separately identifiable profession is being achieved. Most of the present engineeing managers are probably not yet informed about the identification and emergence of this field. Most are not graduates of engineering management programs because these have not been in existence long enough or have been numerous enough to have provided large numbers of graduates. They are not yet widely known to personnel managers and recruiters. It is too early to assess the impact of the newly formed American Society for Engineering Management, but it is expected to be influential in advancing engineering management in theory and practice and maintaining a high professional standard among its members. The existence of this society should in time lead to increased recognition of engineering management as an important profession in the future.

MERRITT A WILLIAMSON
Orrin Henry Ingram Distinguished Professor
of Engineering Management, Emeritus
School of Engineering
Vanderbilt University, Nashville, Tennessee

INFORMATION REFERENCES:

Associations:

    Academy of Management
    American Society for Engineering Education (Engineering Management Division)
    American Society for Engineering Management
    The Institute of Electrical and Electronics Engineers, Inc. (The Engineering Management Society)
    The Institute of Management Sciences (College of Engineering Management)
    Management Committees and Divisions of various Enginering Societies

Periodicals:

    Engineering Management International

Engineering Management Review
R&D Management
Research Management
Transactions on Engineering Management

APPENDIX I

ENGINEERING EDUCATION - AIMS AND GOALS FOR THE EIGHTIES

Franklin Pierce College
Rindge, New Hampshire
July 26-31, 1981

PLANNING COMMITTEE

Chairman: William H. Corcoran, Vice President, ABET
Institute Professor
California Institute of Technology

Co-Chairmen: Leland J. Walker, President, ABET
Chairman of the Board
Northern Testing Laboratories

David R. Reyes-Guerra
Executive Director, ABET

Bruno A. Boley
    Dean, Technological Institute
    Northwestern University

George Burnet
    Coordinator, Engineering Education
    Projects Office
    Iowa State University

Paul S. Chenea
    Vice President, Research Laboratories
    General Motors Corporation

George E. Dieter
    Dean of Engineering
    University of Maryland

Leroy S. Fletcher
    Associate Dean of Engineering
    Texas A&M University

Thomas E. Ford
    Program Director
    Alfred P. Sloan Foundation

John W. Geils
    Director-Network Department Administration
    American Telephone & Telegraph Company

Donald D. Glower
    Dean of Engineering
    Ohio State University

Jerrier A. Haddad
    Vice President, Engineering and Technical
    Personnel Development
    IBM

David C. Hazen
    Executive Director, Assembly of Engineering
    National Research Council

Joseph C. Hogan
    Dean Emeritus, Engineering College
    University of Notre Dame

Thomas R. Horton
    Director, University Relations
    IBM

Samuel F. Hulbert
    President
    Rose-Hulman Institute of Technology

Alfred C. Ingersoll
    College of Engineering
    University of California - Los Angeles

John D. Kemper
    Dean of Engineering
    University of California - Davis

Stothe P. Kezios
    Director, School of Mechanical Engineering
    Georgia Institute of Technolgoy

Irving Leibson
    Vice President
    Bechtel, Inc.

Frank W. Luerssen
    Chairman
    Inland Steel Company

Frank A. McCrackin
    Director, Research and Development
    Southern California Edison Company

Irvan E. Mendenhall
    Daniel, Mann Johnson & Mendenhall

Appendix I

Robert H. Page
    Dean of Engineering
    Texas A&M University

John R. Pierce
    Department of Electrical Engineering
    California Institute of Technology

Robert A. Plane
    President
    Clarkson College

John W. Prados
    Vice President, Academic Affairs
    The University of Tennessee

Peter R. Rony
    Professor, Chemical Engineering
    Virginia Polytechnic Institute and State University

Jack Sanderson
    Assistant Director, Directorate for Engineering
    National Science Foundation

Charles E. Schaffner
    Executive Vice President
    Syska & Hennessey, Inc.

Stuart H. Sherman, Jr.
    Commandant
    Air Force Institute of Technology

Frank B. Sprow
    Vice President
    Exxon Research & Engineering Company

Richard J. Ungrodt
    Vice President Academic Resources and
    Institutional Development
    Milwaukee School of Engineering

APPENDIX II

ENGINEERING EDUCATION - AIMS AND GOALS FOR THE EIGHTIES

Franklin Pierce College
Rindge, New Hampshire
July 26-31, 1981

PARTICIPANTS LIST

Albright, Gifford H.
Professor & Department Head
Architectural Engineering
Pennsylvania State University

Aufill, Charles B.
Continuing Education Manager
Society of Petroleum Engineers

Beckham, Claud L.
Industry Liaison Consultant
Accreditation Board for Engineering and Technology

Bevis, Herbert A.
Associate Dean of Engineering
University of Florida

Boley, Bruno A.
Dean, Technological Institute
Northwestern University

Bordogna, Joseph
Dean, School of Engineering and Applied Science
University of Pennsylvania

Boyle, William R.
Chairman, MERT Division
Oak Ridge Associated University

Browne, Lee F.
Director, Secondary School Relations & Special Student Programs
California Institute of Technology

Corcoran, William H.
Institute Professor
California Institute of Technology

Cole, Sandford S.
Director
Engineering Foundation

Cotton, Jr., Frank E.
    Director, ABET Affairs, AIIE
    Mississippi State University

Crugnola, Aldo M.
    Dean of Engineering
    University of Lowell

Cunningham, Richard G.
    Vice President Research and Graduate Studies
    Pennsylvania State University

Douglas, George W.
    Chairman, Mechanical Engineering Department
    University of South Alabama

Drake, Peter W.
    Assistant Dean of Engineering
    Manhattan College

Eichhorn, Roger
    Dean of Engineering
    University of Kentucky

Ernst, Edward W.
    Professor of Electrical Engineering
    University of Illinois - Urbana

Faiman, Robert N.
    Director of Academic Affairs
    Air Force Institute of Technology

Fischler, Abraham S.
    President
    Nova University

Flammer, Gordon H.
    Professor of Civil Engineering
    Utah State University

Ford, Thomas E.
    Program Director
    Alfred P. Sloan Foundation

Geils, John W.
    Director-Network Department Administration
    American Telephone & Telegraph

Gianniny, Jr., O. Allan
    Associate Professor of Humanities
    College of Engineering
    University of Virginia

Gilliland, Bobby E.
    Associate Dean of Engineering

Clemson University

Glower, Donald D.
   Dean of Engineering
   Ohio State University

Greenfield, Lois B.
   Professor, General Engineering
   University of Wisconsin - Madison

Haddad, Jerrier A.
   Vice President Engineering & Technical
   Personnel Development
   IBM

Haneman, Jr., Vincent S.
   Dean of Engineering
   University of Alaska

Haythornwaite, Robert M.
   Dean of Engineering Technology
   Temple University

Hill, Richard F.
   Dean, College of Science & Engineering
   University of Bridgeport

Hill, Jr., Charles G.
   Professor of Chemical Engineering
   University of Wisconsin - Madison

Hoel, Lester A.
   Chairman, Civil Engineering Department
   University of Virginia

Hogan, Joseph C.
   Dean Emeritus, Engineering College
   University of Notre Dame

Hollander, Lawrence J.
   Associate Dean of Engineering
   The Cooper Union

Hunt, Everett C.
   Head, Engineering Department
   U.S. Merchant Marine Academy

Izadi, Mostafa D.
   Chairman, Civil Engineering Department
   Merrimack College

Jenkins, Ralph E.
   Professor, English Department
   Temple University

Jones, Russel C.
    Dean of Engineering
    University of Massachusetts

Katz, Israel
    Chairman, Engineering Council
    Northeastern University

Kezios, S. Peter
    Director, School of Mechanical Engineering
    Georgia Institute of Technology

Kline, Jacob
    Chairman, Biomedical Engineering
    University of Miami

Kosow, Irving L.
    Professor

Lear, W. Edward
    Executive Director
    American Society for Engineering Education

Lindenlaub, John C.
    Professor, Electrical Engineering
    Purdue University

Manogue, William H.
    Engineering Foundation

Marshall, W. Robert
    Director, University-Industry Research
    University of Wisconsin - Madison

Martin, Ross J.
    Associate Dean of Engineering
    University of Illinois - Urbana

Matthews, Gerald M.
    Director
    Canadian Accreditation Board

Mattson, Roy H.
    Head, Electrical Engineering Department
    University of Arizona

McCollom, Kenneth A.
    Dean of Engineering
    Oklahoma State University

McDevit, William F.
    Vice Chairman, Committee on Educational Aid
    E.I. duPont de Nemours & Company

McDonald, Donald
    Head, Civil Engineering Department
    Texas A&M University

McNamee, Bernard Michael
    Head, Civil Engineering Department
    Drexel University

Morrow, Richard A.
    Chairman, Physics Department
    University of Maine

Mullin, Thomas E.
    Associate Dean, Speed Scientific School
    University of Louisville

Nordby, Gene M.
    Chancellor
    University of Colorado - Denver

O'Neill, Russell R.
    Dean of Engineering
    University of California - Los Angeles

Page, Robert H.
    Dean of Engineering
    Texas A&M University

Parden, Robert J.
    Dean of Engineering
    University of Santa Clara

Plane, Robert Allen
    President
    Clarkson College

Przemieniecki, J.S.
    Dean of Engineering
    Air Force Institute of Technology

Rader, Louis T.
    Professor, Graduate School of Business
    University of Virginia

Reyes-Guerra, David R.
    Executive Director
    Accreditation Board for Engineering and Technology

Richardson, Alfred L.
    Vice President, Corporate Technology
    Pacific Soils Engineering, Inc.

Roadstrum, William H.
    Professor
    Worcester Polytechnic Institute

Roever, Frederick H.

Rony, Peter R.
    Professor, Chemical Engineering
    Virginia Polytechnic Institute and State University

Rosen, Edward M.
    Fellow
    Monsanto Company

Samson, Charles H.
    Acting President
    Texas A&M University

Sanderson Jack
    Assistant Director, Directorate for Engineering
    National Science Foundation

Sangster, William M.
    Dean of Engineering
    Georgia Institute of Technology

Sargent, Jr., Lowrie B
    Senior Scientific Associate
    Alcoa Labs

Schaffner, Charles E.
    Executive Vice President
    Syska & Hennessy, Inc.

Simon, Albert
    Chairman, Mechanical & Aerospace Sciences
    University of Rochester

Sims, James R.
    President-elect, ASCE
    Rice University

Sinclair, George
    Chairman of the Board
    Sinclair Radio Labs, Ltd. (Canada)

Smith, Paul V.
    Manager, Educational Relations
    Exxon Research & Engineering Company

Sprow, Frank B.
    Vice President
    Exxon Research & Engineering Company

Sutton, George E.
    Dean, William Rayen School
    Youngstown State University

Taylor, Harold L.
    Senior Advisor, Research
    Inland Steel Company

Taylor, James I.
    Chairman, Civil Engineering Department
    University of Notre Dame

Thigpen, Jack
    Dean of Engineering
    Louisiana Tech University

Thompson, Hugh A.
    Dean of Engineering
    Tulane University

Ungrodt, Richard J.
    Vice President Academic Resources and Institutional Development
    Milwaukee School of Engineering

Van Valkenburg, M.E.
    Professor, Electrical Engineering
    University of Illinois - Urbana

Vivian, Robert E.
    Dean Emeritus
    University of Southern California

Wakeland, Howard L.
    Associate Dean of Engineering
    University of Illinois - Urbana

Walker, Leland J.
    President, Accreditation Board for Engineering and Technology
    Northern Testing Laboratories

Wasley, Richard J.
    Division Leader
    Lawrence Livermore National Lab

Weinert, Donald G.
    Executive Director
    National Society of Professional Engineers

Wheeler, Orville E.
    Dean, Herff College of Engineering
    Memphis State University

Winkler, Stanley
    IBM

Woodward, James H.
    Dean of Engineering
    University of Alabama

Wyatt, Demarquis D.
　　Staff Officer
　　National Research Council

Zanetti, Peter Carl
　　President
　　Technovate, Inc.

APPENDIX III

ENGINEERING EDUCATION - AIMS AND GOALS FOR THE EIGHTIES

An Engineering Foundation Conference

Sponsored by

Engineering Foundation
ACCREDITATION BOARD FOR ENGINEERING AND TECHNOLOGY

Co-sponsored by

National Academy of Engineering
American Society for Engineering Education
Educational Affairs Council of AAES

Franklin Pierce College
Rindge, New Hampshire
July 26-31, 1981

SCHEDULE OF MAIN SESSIONS
AND SESSION LEADERS

MONDAY, JULY 27

Session #

(1) Cooperation among scientists, engineers, universities and industry in on-going development of high school curricula

    10 am-11 am   JOSEPH BORDOGNA

(2) Role of professional societies in high school counseling and value of organizations such as JETS, NECG, EdAC, Boy Scouts of America

    11 am-12 noon   JACK GEILS

(3) Should engineering schools and industry be involved in development of pre-college curricula?

    7 pm-8 pm   WILLIAM F. McDEVIT

(4) Opportunities for the gifted student with engineering interests

    8 pm-9 pm   LEE F. BROWNE

(5) Improved communication with minority students on opportunities in the engineering profession

9 pm-10pm   DAVID R. REYES-GUERRA

TUESDAY, JULY 28

(6) Engineering education - the liberal education of the eighties

9 am-10am   O. ALLAN GIANNINY

(7) Increased interaction of scientists and engineers in development of both engineering and science curricula

10 am-11am   WILLIAM H. CORCORAN

(8) Effect of progress in microelectronics in education in electrical engineering especially and engineering in general

11 am-12noon   LOUIS RADER

(9) Teaching of engineering courses with student faculty ratios far in excess of past experiences

7 pm-8pm   JOHN C. LINDENLAUB

(10) Development of clear, easy writing and clear, plain speaking

8 pm-9pm   RALPH JENKINS

(11) Key aspects of comparison between engineering education in USA and England, Japan, Germany and Russia

9 pm-10pm   J.S. PRZEMIENIECKI

WEDNESDAY, JULY 29

(12) How to increase the number of Ph.D. candidates and supply of faculty members

9 am-10am   JERRIER HADDAD and DONALD GLOWER

(13) Implications of increasing percentage of foreign nationals enrolled in graduate programs for engineering

10 am-11am   W. ROBERT MARSHALL

(14) Effect of Federal support for programs and students on generation of new faculty members

11 am-12noon   ROSS MARTIN

(15) Effect of Federal support on research by faculty and graduate students

7 pm-8pm   RICHARD G. CUNNINGHAM and
                    JACK SANDERSON

Appendix III

(16) The Master's degree as the entry level for engineering practice

    8 pm-9 pm   GENE M. NORDBY

(17) Concept of a professional school engineering

    9 pm-10 pm   CHARLES SAMSON

THURSDAY, JULY 30

(18) Need for continuing education and control of quality by the university, the professional society, the industrial organization and the proprietary school

    9 am- 10 am   ROY MATTSON

(19) Effect of computer graphics on both industry and engineering education

    10 am-11 am   EDWARD M. ROSEN

(20) Implications (such as what is lacking and how it might be handled) for industry and for engineering education of increased use of science graduates in engineering roles

    11 am-12 noon   DEMARQUIS WYATT

(21) Effect on industrial design and on engineering education of desk computers joinable to networks

    7 pm- 8pm   PETER RONY

(22) Roles of industry, advisory committees, and government in changing engineering education

    8 pm-9 pm   S. PETER KEZIOS

(23) Participation by industry in education of needed engineers

    9 pm-10 pm   FRANK B. SPROW

FRIDAY, JULY 31

(24) CONSENSUS VIEWS INCLUDING RECOMMENDATIONS FOR FOLLOW-UP

    24.1   The pre-college experience for engineering education

           9 am-9:30 am   DAVID R. REYES-GUERRA

    24.2   Engineering education and the baccalaureate experience

           9:30 am-10 am   RUSSELL R. O'NEILL

    24.3   Graduate education in engineering

ENGINEERING EDUCATION

        10 am-10:30am   ROBERT H. PAGE

24.4 Engineering practice and engineering education

        10:30am-11am   CHARLES SCHAFFNER

24.5 Research and Engineering Education

        11 am-11:30 am   RICHARD G. CUNNINGHAM

24.6 Communication of conference results to President Reagan

        11:30 am-12 noon   LELAND J. WALKER

Author Index

Browne, Lee F.
An Examination and Discussion of Opportunities Available to Gifted Students with Engineering Interests — Problems and Interpretations, 41

Cunningham, Richard C.
Summary and Recommendations: Research and Engineering Education, 35

Gianniny, D. Allan
Engineering Education and the Baccalaureate Experience (Engineering Education — The Liberal Education of the 80's), 51

Glower, Donald D.
How to Increase the Number of Ph.D. Candidates and the Supply of New Faculty Members, 97

Haddad, Jerrier A.
Crisis in American Engineering Education, 89

Jenkins, Ralph E.
Why are Engineering Students Not Being Taught to Write and Speak Well, and What are We Going to Do About It?, 69

Lindenlaub, John C.
Teaching of Engineering Courses with Student Faculty Ratios Far in Excess of Past Experience, 65

Marshall, W. Robert
Implications of Increasing Percentage of Foreign Nationals Enrolled in Graduate Programs for Engineering, 103

Martin, Ross J.
Effect of Federal Support for Programs and Students on Generation of New Faculty, 109

Matson, Roy H.
The Need for Continuing Education and Its Quality Control, 137

O'Neill, Russell R.
Summary and Recommendations: Engineering Education and the Baccalaureate Experience, 21

Page, Robert H.
Summary and Recommendations: Graduate Education in Engineering, 25

Przemieniecki, J. S.
Key Aspects of Comparison Between Engineering Education in the USA and England, Japan, Germany, and Russia, 77

Rader, L. T.
Effect of Progress in Microelectronics in Education, in Electrical Engineering Especially, and Engineering Education in General, 53

Reyes-Guerra, David R.
Summary and Recommendations: The Pre-College Experience for Engineering Education, 19

Rosen, Edward M.
Effect of Computer Graphics in Both Industry and Engineering Education, 139

Samson, Charles H.
Concept of a Professional School of Engineering, 117

Schaffner, Charles E.
Summary and Recommendations: Engineering Practice and Engineering Education, 29

Walker, Leland J.
Opening Remarks, 39

Williamson, Merritt A.
Education for Engineering Management in the Eighties, 157

Subject Index

Accreditation; Computers; Computer terminals; Corporate responsibility; Education-practice interchange; Electrical engineering; Engineering education; Laboratory equipment; Lasers; Microelectronics; Productivity; Research
  Effect of Progress in Microelectronics in Education, in Electrical Engineering Especially, and Engineering Education in General, L. T. Rader, 53

Accreditation; Continuing education; Education-practice interchange; Industries; Licensing; Motivation; Professional engineering; Quality control; Social aspects; Sociological factors; Validation
  The Need for Continuing Education and Its Quality Control, Roy H. Matson, 137

Analysis; Comprehension; Education-practice interchange; Engineering schools; Evaluation; Graduate study; Logistics; Students; Synthesis; Undergraduate study
  Teaching of Engineering Courses with Student Faculty Ratios Far in Excess of Past Experience, John C. Lindenlaub, 65

Communication between engineers; Computer applications; Curricula; Engineering education; Humanities; Sciences; Students; Teaching methods and content; Undergraduate study
  Summary and Recommendations: Engineering Education and the Baccalaureate Experience, Russell R. O'Neill, 21

Communication between engineers; Curricula; Engineering schools; Human factors; Information; Languages; Research; Technical writers; Writing
  Why are Engineering Students Not Being Taught to Write and Speak Well, and What are We Going to Do About It?, Ralph E. Jenkins, 69

Comprehension; Education-practice interchange; Engineering schools; Evaluation; Graduate study; Logistics; Students; Synthesis; Undergraduate study; Analysis
  Teaching of Engineering Courses with Student Faculty Ratios Far in Excess of Past Experience, John C. Lindenlaub, 65

Computer applications; Computer graphics; Continuing education; Engineering education; Federal aid; Graduate study; Industries; Research; Salaries
  Summary and Recommendations: Graduate Education in Engineering, Robert H. Page, 25

Computer applications; Curricula; Engineering education; Humanities; Sciences; Students; Teaching methods and content; Undergraduate study; Communication between engineers
  Summary and Recommendations: Engineering Education and the Baccalaureate Experience, Russell R. O'Neill, 21

Computer graphics; Computers; Cost effectiveness; Engineering education; Engineering schools; Industries; Man machine systems; Productivity; Sciences
  Effect of Computer Graphics in Both Industry and Engineering Education, Edward M. Rosen, 139

Computer graphics; Continuing education; Engineering education; Federal aid; Graduate study; Industries; Research; Salaries; Computer applications
  Summary and Recommendations: Graduate Education in Engineering, Robert H. Page, 25

Computers; Computer terminals; Corporate responsibility; Education-practice interchange; Electrical engineering; Engineering education; Laboratory equipment; Lasers; Microelectronics; Productivity; Research; Accreditation
  Effect of Progress in Microelectronics in Education, in Electrical Engineering Especially, and Engineering Education in General, L. T. Rader, 53

Computers; Cost effectiveness; Engineering education; Engineering schools; Industries; Man machine systems; Productivity; Sciences; Computer graphics
  Effect of Computer Graphics in Both Industry and Engineering Education, Edward M. Rosen, 139

Computer terminals; Corporate responsibility; Education-practice interchange; Electrical engineering; Engineering education; Laboratory equipment; Lasers; Microelectronics; Productivity; Research; Accreditation; Computers
  Effect of Progress in Microelectronics in Education, in Electrical Engineering Especially, and Engineering Education in General, L. T. Rader, 53

Consulting engineers; Curricula; Education-practice interchange; Engineering education; Engineers; Model studies; Professional development; Professional engineering; Professional schools; Research
  Concept of a Professional School of Engineering, Charles H. Samson, 117

Continuing education; Education-practice interchange; Engineering; Engineering education; Graduate study; Industries; Minority groups; Professional engineering; Salaries
  Summary and Recommendations: Engineering Practice and Engineering Education, Charles E. Schaffner, 29

Continuing education; Education-practice interchange; Industries; Licensing; Motivation; Professional engineering; Quality control; Social aspects; Sociological factors; Validation; Accreditation
  The Need for Continuing Education and Its Quality Control, Roy H. Matson, 137

Continuing education; Engineering education; Federal aid; Graduate study; Industries; Research; Salaries; Computer applications; Computer graphics
  Summary and Recommendations: Graduate Education in Engineering, Robert H. Page, 25

ENGINEERING EDUCATION

Corporate responsibility; Education-practice interchange; Electrical engineering; Engineering education; Laboratory equipment; Lasers; Microelectronics; Productivity; Research; Accreditation; Computers; Computer terminals
  Effect of Progress in Microelectronics in Education, in Electrical Engineering Especially, and Engineering Education in General, L. T. Rader, 53

Cost effectiveness; Engineering education; Engineering schools; Industries; Man machine systems; Productivity; Sciences; Computer graphics; Computers
  Effect of Computer Graphics in Both Industry and Engineering Education, Edward M. Rosen, 139

Curricula; Education-practice interchange; Engineering education; Engineers; Model studies; Professional development; Professional engineering; Professional schools; Research; Consulting engineers
  Concept of a Professional School of Engineering, Charles H. Samson, 117

Curricula; Engineering; Forecasting; Management; Management engineering; Management systems; Management training; Professional engineering; Universities
  Education for Engineering Management in the Eighties, Merritt A. Williamson, 157

Curricula; Engineering education; Germany; Graduate study; Japan; Union of Soviet Socialist Republics; United Kingdom; United States
  Key Aspects of Comparison Between Engineering Education in the USA and England, Japan, Germany, and Russia, J. S. Przemieniecki, 77

Curricula; Engineering education; Human factors; Humanities; Mathematics; Sciences; Social sciences; Undergraduate study
  Engineering Education and the Baccalaureate Experience (Engineering Education — The Liberal Education of the 80's), D. Allan Gianniny, 51

Curricula; Engineering education; Humanities; Sciences; Students; Teaching methods and content; Undergraduate study; Communication between engineers; Computer applications
  Summary and Recommendations: Engineering Education and the Baccalaureate Experience, Russell R. O'Neill, 21

Curricula; Engineering schools; Human factors; Information; Languages; Research; Technical writers; Writing; Communication between engineers
  Why are Engineering Students Not Being Taught to Write and Speak Well, and What are We Going to Do About It?, Ralph E. Jenkins, 69

Demand; Engineering education; Government; Industries; Mathematics; Minority groups; Motivation; Salaries; Sciences
  Summary and Recommendations: The Pre-College Experience for Engineering Education, David R. Reyes-Guerra, 19

Economic factors; Engineering education; Federal aid; Financing; Funding allocations; Graduate study; Laboratory equipment; Productivity; Recruiting; Universities
  How to Increase the Number of Ph.D. Candidates and the Supply of New Faculty Members, Donald D. Glower, 97

Education-practice interchange; Electrical engineering; Engineering education; Laboratory equipment; Lasers; Microelectronics; Productivity; Research; Accreditation; Computers; Computer terminals; Corporate responsibility
  Effect of Progress in Microelectronics in Education, in Electrical Engineering Especially, and Engineering Education in General, L. T. Rader, 53

Education-practice interchange; Engineering; Engineering education; Graduate study; Industries; Minority groups; Professional engineering; Salaries; Continuing education
  Summary and Recommendations: Engineering Practice and Engineering Education, Charles E. Schaffner, 29

Education-practice interchange; Engineering education; Engineers; Model studies; Professional development; Professional engineering; Professional schools; Research; Consulting engineers; Curricula
  Concept of a Professional School of Engineering, Charles H. Samson, 117

Education-practice interchange; Engineering education; Federal aid; Graduate study; Industries; Laboratory equipment; Research; Undergraduate study
  Summary and Recommendations: Research and Engineering Education, Richard C. Cunningham, 35

Education-practice interchange; Engineering schools; Evaluation; Graduate study; Logistics; Students; Synthesis; Undergraduate study; Analysis; Comprehension
  Teaching of Engineering Courses with Student Faculty Ratios Far in Excess of Past Experience, John C. Lindenlaub, 65

Education-practice interchange; Industries; Licensing; Motivation; Professional engineering; Quality control; Social aspects; Sociological factors; Validation; Accreditation; Continuing education
  The Need for Continuing Education and Its Quality Control, Roy H. Matson, 137

Electrical engineering; Engineering education; Laboratory equipment; Lasers; Microelectronics; Productivity; Research; Accreditation; Computers; Computer terminals; Corporate responsibility; Education-practice interchange
  Effect of Progress in Microelectronics in Education, in Electrical Engineering Especially, and Engineering Education in General, L. T. Rader, 53

Engineering; Engineering education; Engineering schools; Financing; Government; Industries; Laboratory equipment; Recruiting
  Opening Remarks, Leland J. Walker, 39

Engineering; Engineering education; Engineering schools; Minority groups; Research; Students; Women
  An Examination and Discussion of Opportunities Available to Gifted Students with Engineering Interests — Problems and Interpretations, Lee F. Browne, 41

Subject Index

Engineering; Engineering education; Graduate study; Industries; Minority groups; Professional engineering; Salaries; Continuing education; Education-practice interchange
Summary and Recommendations: Engineering Practice and Engineering Education, Charles E. Schaffner, 29

Engineering; Forecasting; Management; Management engineering; Management systems; Management training; Professional engineering; Universities; Curricula
Education for Engineering Management in the Eighties, Merritt A. Williamson, 157

Engineering education; Engineering schools; Financing; Government; Industries; Laboratory equipment; Recruiting; Engineering
Opening Remarks, Leland J. Walker, 39

Engineering education; Engineering schools; Industries; Man machine systems; Productivity; Sciences; Computer graphics; Computers; Cost effectiveness
Effect of Computer Graphics in Both Industry and Engineering Education, Edward M. Rosen, 139

Engineering education; Engineering schools; Minority groups; Research; Students; Women; Engineering
An Examination and Discussion of Opportunities Available to Gifted Students with Engineering Interests — Problems and Interpretations, Lee F. Browne, 41

Engineering education; Engineers; Model studies; Professional development; Professional engineering; Professional schools; Research; Consulting engineers; Curricula; Education-practice interchange
Concept of a Professional School of Engineering, Charles H. Samson, 117

Engineering education; Federal aid; Financing; Funding allocations; Graduate study; Laboratory equipment; Productivity; Recruiting; Universities; Economic factors
How to Increase the Number of Ph.D. Candidates and the Supply of New Faculty Members, Donald D. Glower, 97

Engineering education; Federal aid; Graduate study; Industries; Laboratory equipment; Research; Statistical data; Statistical distributions; Students; Teaching methods and content
Effect of Federal Support for Programs and Students on Generation of New Faculty, Ross J. Martin, 109

Engineering education; Federal aid; Graduate study; Industries; Laboratory equipment; Research; Undergraduate study; Education-practice interchange
Summary and Recommendations: Research and Engineering Education, Richard C. Cunningham, 35

Engineering education; Federal aid; Graduate study; Industries; Research; Salaries; Computer applications; Computer graphics; Continuing education
Summary and Recommendations: Graduate Education in Engineering, Robert H. Page, 25

Engineering education; Foreign engineering; Graduate study; Research; Statistical distributions; Students; Undergraduate study; Universities
Implications of Increasing Percentage of Foreign Nationals Enrolled in Graduate Programs for Engineering, W. Robert Marshall, 103

Engineering education; Germany; Graduate study; Japan; Union of Soviet Socialist Republics; United Kingdom; United States; Curricula
Key Aspects of Comparison Between Engineering Education in the USA and England, Japan, Germany, and Russia, J. S. Przemieniecki, 77

Engineering education; Government; Graduate study; Industries; Inflation (economic); Patents; Productivity; Salaries; United States; Universities
Crisis in American Engineering Education, Jerrier A. Haddad, 89

Engineering education; Government; Industries; Mathematics; Minority groups; Motivation; Salaries; Sciences; Demand
Summary and Recommendations: The Pre-College Experience for Engineering Education, David R. Reyes-Guerra, 19

Engineering education; Graduate study; Industries; Minority groups; Professional engineering; Salaries; Continuing education; Education-practice interchange; Engineering
Summary and Recommendations: Engineering Practice and Engineering Education, Charles E. Schaffner, 29

Engineering education; Human factors; Humanities; Mathematics; Sciences; Social sciences; Undergraduate study; Curricula
Engineering Education and the Baccalaureate Experience (Engineering Education — The Liberal Education of the 80's), D. Allan Gianniny, 51

Engineering education; Humanities; Sciences; Students; Teaching methods and content; Undergraduate study; Communication between engineers; Computer applications; Curricula
Summary and Recommendations: Engineering Education and the Baccalaureate Experience, Russell R. O'Neill, 21

Engineering education; Laboratory equipment; Lasers; Microelectronics; Productivity; Research; Accreditation; Computers; Computer terminals; Corporate responsibility; Education-practice interchange; Electrical engineering
Effect of Progress in Microelectronics in Education, in Electrical Engineering Especially, and Engineering Education in General, L. T. Rader, 53

Engineering schools; Evaluation; Graduate study; Logistics; Students; Synthesis; Undergraduate study; Analysis; Comprehension; Education-practice interchange
Teaching of Engineering Courses with Student Faculty Ratios Far in Excess of Past Experience, John C. Lindenlaub, 65

Engineering schools; Financing; Government; Industries; Laboratory equipment; Recruiting; Engineering; Engineering education
Opening Remarks, Leland J. Walker, 39

Engineering schools; Human factors; Information; Languages; Research; Technical writers; Writing; Communication between engineers; Curricula
  Why are Engineering Students Not Being Taught to Write and Speak Well, and What are We Going to Do About It?, Ralph E. Jenkins, 69

Engineering schools; Industries; Man machine systems; Productivity; Sciences; Computer graphics; Computers; Cost effectiveness; Engineering education
  Effect of Computer Graphics in Both Industry and Engineering Education, Edward M. Rosen, 139

Engineering schools; Minority groups; Research; Students; Women; Engineering; Engineering education
  An Examination and Discussion of Opportunities Available to Gifted Students with Engineering Interests — Problems and Interpretations, Lee F. Browne, 41

Engineers; Model studies; Professional development; Professional engineering; Professional schools; Research; Consulting engineers; Curricula; Education-practice interchange; Engineering education
  Concept of a Professional School of Engineering, Charles H. Samson, 117

Evaluation; Graduate study; Logistics; Students; Synthesis; Undergraduate study; Analysis; Comprehension; Education-practice interchange; Engineering schools
  Teaching of Engineering Courses with Student Faculty Ratios Far in Excess of Past Experience, John C. Lindenlaub, 65

Federal aid; Financing; Funding allocations; Graduate study; Laboratory equipment; Productivity; Recruiting; Universities; Economic factors; Engineering education
  How to Increase the Number of Ph.D. Candidates and the Supply of New Faculty Members, Donald D. Glower, 97

Federal aid; Graduate study; Industries; Laboratory equipment; Research; Statistical data; Statistical distributions; Students; Teaching methods and content; Engineering education
  Effect of Federal Support for Programs and Students on Generation of New Faculty, Ross J. Martin, 109

Federal aid; Graduate study; Industries; Laboratory equipment; Research; Undergraduate study; Education-practice interchange; Engineering education
  Summary and Recommendations: Research and Engineering Education, Richard C. Cunningham, 35

Federal aid; Graduate study; Industries; Research; Salaries; Computer applications; Computer graphics; Continuing education; Engineering education
  Summary and Recommendations: Graduate Education in Engineering, Robert H. Page, 25

Financing; Funding allocations; Graduate study; Laboratory equipment; Productivity; Recruiting; Universities; Economic factors; Engineering education; Federal aid
  How to Increase the Number of Ph.D. Candidates and the Supply of New Faculty Members, Donald D. Glower, 97

Financing; Government; Industries; Laboratory equipment; Recruiting; Engineering; Engineering education; Engineering schools
  Opening Remarks, Leland J. Walker, 39

Forecasting; Management; Management engineering; Management systems; Management training; Professional engineering; Universities; Curricula; Engineering
  Education for Engineering Management in the Eighties, Merritt A. Williamson, 157

Foreign engineering; Graduate study; Research; Statistical distributions; Students; Undergraduate study; Universities; Engineering education
  Implications of Increasing Percentage of Foreign Nationals Enrolled in Graduate Programs for Engineering, W. Robert Marshall, 103

Funding allocations; Graduate study; Laboratory equipment; Productivity; Recruiting; Universities; Economic factors; Engineering education; Federal aid; Financing
  How to Increase the Number of Ph.D. Candidates and the Supply of New Faculty Members, Donald D. Glower, 97

Germany; Graduate study; Japan; Union of Soviet Socialist Republics; United Kingdom; United States; Curricula; Engineering education
  Key Aspects of Comparison Between Engineering Education in the USA and England, Japan, Germany, and Russia, J. S. Przemieniecki, 77

Government; Graduate study; Industries; Inflation (economic); Patents; Productivity; Salaries; United States; Universities; Engineering education
  Crisis in American Engineering Education, Jerrier A. Haddad, 89

Government; Industries; Laboratory equipment; Recruiting; Engineering; Engineering education; Engineering schools; Financing
  Opening Remarks, Leland J. Walker, 39

Government; Industries; Mathematics; Minority groups; Motivation; Salaries; Sciences; Demand; Engineering education
  Summary and Recommendations: The Pre-College Experience for Engineering Education, David R. Reyes-Guerra, 19

Graduate study; Industries; Inflation (economic); Patents; Productivity; Salaries; United States; Universities; Engineering education; Government
  Crisis in American Engineering Education, Jerrier A. Haddad, 89

Graduate study; Industries; Laboratory equipment; Research; Statistical data; Statistical distributions; Students; Teaching methods and content; Engineering education; Federal aid
  Effect of Federal Support for Programs and Students on Generation of New Faculty, Ross J. Martin, 109

Graduate study; Industries; Laboratory equipment; Research; Undergraduate study; Education-practice interchange; Engineering education; Federal aid
  Summary and Recommendations: Research and Engineering Education, Richard C. Cunningham, 35

Subject Index

Graduate study; Industries; Minority groups; Professional engineering; Salaries; Continuing education; Education-practice interchange; Engineering; Engineering education
Summary and Recommendations: Engineering Practice and Engineering Education, Charles E. Schaffner, 29

Graduate study; Industries; Research; Salaries; Computer applications; Computer graphics; Continuing education; Engineering education; Federal aid
Summary and Recommendations: Graduate Education in Engineering, Robert H. Page, 25

Graduate study; Japan; Union of Soviet Socialist Republics; United Kingdom; United States; Curricula; Engineering education; Germany
Key Aspects of Comparison Between Engineering Education in the USA and England, Japan, Germany, and Russia, J. S. Przemieniecki, 77

Graduate study; Laboratory equipment; Productivity; Recruiting; Universities; Economic factors; Engineering education; Federal aid; Financing; Funding allocations
How to Increase the Number of Ph.D. Candidates and the Supply of New Faculty Members, Donald D. Glower, 97

Graduate study; Logistics; Students; Synthesis; Undergraduate study; Analysis; Comprehension; Education-practice interchange; Engineering schools; Evaluation
Teaching of Engineering Courses with Student Faculty Ratios Far in Excess of Past Experience, John C. Lindenlaub, 65

Graduate study; Research; Statistical distributions; Students; Undergraduate study; Universities; Engineering education; Foreign engineering
Implications of Increasing Percentage of Foreign Nationals Enrolled in Graduate Programs for Engineering, W. Robert Marshall, 103

Human factors; Humanities; Mathematics; Sciences; Social sciences; Undergraduate study; Curricula; Engineering education
Engineering Education and the Baccalaureate Experience (Engineering Education — The Liberal Education of the 80's), D. Allan Gianniny, 51

Human factors; Information; Languages; Research; Technical writers; Writing; Communication between engineers; Curricula; Engineering schools
Why are Engineering Students Not Being Taught to Write and Speak Well, and What are We Going to Do About It?, Ralph E. Jenkins, 69

Humanities; Mathematics; Sciences; Social sciences; Undergraduate study; Curricula; Engineering education; Human factors
Engineering Education and the Baccalaureate Experience (Engineering Education — The Liberal Education of the 80's), D. Allan Gianniny, 51

Humanities; Sciences; Students; Teaching methods and content; Undergraduate study; Communication between engineers; Computer applications; Curricula; Engineering education
Summary and Recommendations: Engineering Education and the Baccalaureate Experience, Russell R. O'Neill, 21

Industries; Inflation (economic); Patents; Productivity; Salaries; United States; Universities; Engineering education; Government; Graduate study
Crisis in American Engineering Education, Jerrier A. Haddad, 89

Industries; Laboratory equipment; Recruiting; Engineering; Engineering education; Engineering schools; Financing; Government
Opening Remarks, Leland J. Walker, 39

Industries; Laboratory equipment; Research; Statistical data; Statistical distributions; Students; Teaching methods and content; Engineering education; Federal aid; Graduate study
Effect of Federal Support for Programs and Students on Generation of New Faculty, Ross J. Martin, 109

Industries; Laboratory equipment; Research; Undergraduate study; Education-practice interchange; Engineering education; Federal aid; Graduate study
Summary and Recommendations: Research and Engineering Education, Richard C. Cunningham, 35

Industries; Licensing; Motivation; Professional engineering; Quality control; Social aspects; Sociological factors; Validation; Accreditation; Continuing education; Education-practice interchange
The Need for Continuing Education and Its Quality Control, Roy H. Matson, 137

Industries; Man machine systems; Productivity; Sciences; Computer graphics; Computers; Cost effectiveness; Engineering education; Engineering schools
Effect of Computer Graphics in Both Industry and Engineering Education, Edward M. Rosen, 139

Industries; Mathematics; Minority groups; Motivation; Salaries; Sciences; Demand; Engineering education; Government
Summary and Recommendations: The Pre-College Experience for Engineering Education, David R. Reyes-Guerra, 19

Industries; Minority groups; Professional engineering; Salaries; Continuing education; Education-practice interchange; Engineering; Engineering education; Graduate study
Summary and Recommendations: Engineering Practice and Engineering Education, Charles E. Schaffner, 29

Industries; Research; Salaries; Computer applications; Computer graphics; Continuing education; Engineering education; Federal aid; Graduate study
Summary and Recommendations: Graduate Education in Engineering, Robert H. Page, 25

Inflation (economic); Patents; Productivity; Salaries; United States; Universities; Engineering education; Government; Graduate study; Industries
Crisis in American Engineering Education, Jerrier A. Haddad, 89

Information; Languages; Research; Technical writers; Writing; Communication between engineers; Curricula; Engineering schools; Human factors
Why are Engineering Students Not Being Taught to Write and Speak Well, and What are We Going to Do About It?, Ralph E. Jenkins, 69

Japan; Union of Soviet Socialist Republics; United Kingdom; United States; Curricula; Engineering education; Germany; Graduate study
   Key Aspects of Comparison Between Engineering Education in the USA and England, Japan, Germany, and Russia, J. S. Przemieniecki, 77

Laboratory equipment; Lasers; Microelectronics; Productivity; Research; Accreditation; Computers; Computer terminals; Corporate responsibility; Education-practice interchange; Electrical engineering; Engineering education
   Effect of Progress in Microelectronics in Education, in Electrical Engineering Especially, and Engineering Education in General, L. T. Rader, 53

Laboratory equipment; Productivity; Recruiting; Universities; Economic factors; Engineering education; Federal aid; Financing; Funding allocations; Graduate study
   How to Increase the Number of Ph.D. Candidates and the Supply of New Faculty Members, Donald D. Glower, 97

Laboratory equipment; Recruiting; Engineering; Engineering education; Engineering schools; Financing; Government; Industries
   Opening Remarks, Leland J. Walker, 39

Laboratory equipment; Research; Statistical data; Statistical distributions; Students; Teaching methods and content; Engineering education; Federal aid; Graduate study; Industries
   Effect of Federal Support for Programs and Students on Generation of New Faculty, Ross J. Martin, 109

Laboratory equipment; Research; Undergraduate study; Education-practice interchange; Engineering education; Federal aid; Graduate study; Industries
   Summary and Recommendations: Research and Engineering Education, Richard C. Cunningham, 35

Languages; Research; Technical writers; Writing; Communication between engineers; Curricula; Engineering schools; Human factors; Information
   Why are Engineering Students Not Being Taught to Write and Speak Well, and What are We Going to Do About It?, Ralph E. Jenkins, 69

Lasers; Microelectronics; Productivity; Research; Accreditation; Computers; Computer terminals; Corporate responsibility; Education-practice interchange; Electrical engineering; Engineering education; Laboratory equipment
   Effect of Progress in Microelectronics in Education, in Electrical Engineering Especially, and Engineering Education in General, L. T. Rader, 53

Licensing; Motivation; Professional engineering; Quality control; Social aspects; Sociological factors; Validation; Accreditation; Continuing education; Education-practice interchange; Industries
   The Need for Continuing Education and Its Quality Control, Roy H. Matson, 137

Logistics; Students; Synthesis; Undergraduate study; Analysis; Comprehension; Education-practice interchange; Engineering schools; Evaluation; Graduate study
   Teaching of Engineering Courses with Student Faculty Ratios Far in Excess of Past Experience, John C. Lindenlaub, 65

Management; Management engineering; Management systems; Management training; Professional engineering; Universities; Curricula; Engineering; Forecasting
   Education for Engineering Management in the Eighties, Merritt A. Williamson, 157

Management engineering; Management systems; Management training; Professional engineering; Universities; Curricula; Engineering; Forecasting; Management
   Education for Engineering Management in the Eighties, Merritt A. Williamson, 157

Management systems; Management training; Professional engineering; Universities; Curricula; Engineering; Forecasting; Management; Management engineering
   Education for Engineering Management in the Eighties, Merritt A. Williamson, 157

Management training; Professional engineering; Universities; Curricula; Engineering; Forecasting; Management; Management engineering; Management systems
   Education for Engineering Management in the Eighties, Merritt A. Williamson, 157

Man machine systems; Productivity; Sciences; Computer graphics; Computers; Cost effectiveness; Engineering education; Engineering schools; Industries
   Effect of Computer Graphics in Both Industry and Engineering Education, Edward M. Rosen, 139

Mathematics; Minority groups; Motivation; Salaries; Sciences; Demand; Engineering education; Government; Industries
   Summary and Recommendations: The Pre-College Experience for Engineering Education, David R. Reyes-Guerra, 19

Mathematics; Sciences; Social sciences; Undergraduate study; Curricula; Engineering education; Human factors; Humanities
   Engineering Education and the Baccalaureate Experience (Engineering Education — The Liberal Education of the 80's), D. Allan Gianniny, 51

Microelectronics; Productivity; Research; Accreditation; Computers; Computer terminals; Corporate responsibility; Education-practice interchange; Electrical engineering; Engineering education; Laboratory equipment; Lasers
   Effect of Progress in Microelectronics in Education, in Electrical Engineering Especially, and Engineering Education in General, L. T. Rader, 53

Minority groups; Motivation; Salaries; Sciences; Demand; Engineering education; Government; Industries; Mathematics
   Summary and Recommendations: The Pre-College Experience for Engineering Education, David R. Reyes-Guerra, 19

Minority groups; Professional engineering; Salaries; Continuing education; Education-practice interchange; Engineering; Engineering education; Graduate study; Industries
   Summary and Recommendations: Engineering Practice and Engineering Education, Charles E. Schaffner, 29

Subject Index

Minority groups; Research; Students; Women; Engineering; Engineering education; Engineering schools
An Examination and Discussion of Opportunities Available to Gifted Students with Engineering Interests — Problems and Interpretations, Lee F. Browne, 41

Model studies; Professional development; Professional engineering; Professional schools; Research; Consulting engineers; Curricula; Education-practice interchange; Engineering education; Engineers
Concept of a Professional School of Engineering, Charles H. Samson, 117

Motivation; Professional engineering; Quality control; Social aspects; Sociological factors; Validation; Accreditation; Continuing education; Education-practice interchange; Industries; Licensing
The Need for Continuing Education and Its Quality Control, Roy H. Matson, 137

Motivation; Salaries; Sciences; Demand; Engineering education; Government; Industries; Mathematics; Minority groups
Summary and Recommendations: The Pre-College Experience for Engineering Education, David R. Reyes-Guerra, 19

Patents; Productivity; Salaries; United States; Universities; Engineering education; Government; Graduate study; Industries; Inflation (economic)
Crisis in American Engineering Education, Jerrier A. Haddad, 89

Productivity; Recruiting; Universities; Economic factors; Engineering education; Federal aid; Financing; Funding allocations; Graduate study; Laboratory equipment
How to Increase the Number of Ph.D. Candidates and the Supply of New Faculty Members, Donald D. Glower, 97

Productivity; Research; Accreditation; Computers; Computer terminals; Corporate responsibility; Education-practice interchange; Electrical engineering; Engineering education; Laboratory equipment; Lasers; Microelectronics
Effect of Progress in Microelectronics in Education, in Electrical Engineering Especially, and Engineering Education in General, L. T. Rader, 53

Productivity; Salaries; United States; Universities; Engineering education; Government; Graduate study; Industries; Inflation (economic); Patents
Crisis in American Engineering Education, Jerrier A. Haddad, 89

Productivity; Sciences; Computer graphics; Computers; Cost effectiveness; Engineering education; Engineering schools; Industries; Man machine systems
Effect of Computer Graphics in Both Industry and Engineering Education, Edward M. Rosen, 139

Professional development; Professional engineering; Professional schools; Research; Consulting engineers; Curricula; Education-practice interchange; Engineering education; Engineers; Model studies
Concept of a Professional School of Engineering, Charles H. Samson, 117

Professional engineering; Professional schools; Research; Consulting engineers; Curricula; Education-practice interchange; Engineering education; Engineers; Model studies; Professional development
Concept of a Professional School of Engineering, Charles H. Samson, 117

Professional engineering; Quality control; Social aspects; Sociological factors; Validation; Accreditation; Continuing education; Education-practice interchange; Industries; Licensing; Motivation
The Need for Continuing Education and Its Quality Control, Roy H. Matson, 137

Professional engineering; Salaries; Continuing education; Education-practice interchange; Engineering; Engineering education; Graduate study; Industries; Minority groups
Summary and Recommendations: Engineering Practice and Engineering Education, Charles E. Schaffner, 29

Professional engineering; Universities; Curricula; Engineering; Forecasting; Management; Management engineering; Management systems; Management training
Education for Engineering Management in the Eighties, Merritt A. Williamson, 157

Professional schools; Research; Consulting engineers; Curricula; Education-practice interchange; Engineering education; Engineers; Model studies; Professional development; Professional engineering
Concept of a Professional School of Engineering, Charles H. Samson, 117

Quality control; Social aspects; Sociological factors; Validation; Accreditation; Continuing education; Education-practice interchange; Industries; Licensing; Motivation; Professional engineering
The Need for Continuing Education and Its Quality Control, Roy H. Matson, 137

Recruiting; Engineering; Engineering education; Engineering schools; Financing; Government; Industries; Laboratory equipment
Opening Remarks, Leland J. Walker, 39

Recruiting; Universities; Economic factors; Engineering education; Federal aid; Financing; Funding allocations; Graduate study; Laboratory equipment; Productivity
How to Increase the Number of Ph.D. Candidates and the Supply of New Faculty Members, Donald D. Glower, 97

Research; Accreditation; Computers; Computer terminals; Corporate responsibility; Education-practice interchange; Electrical engineering; Engineering education; Laboratory equipment; Lasers; Microelectronics; Productivity
Effect of Progress in Microelectronics in Education, in Electrical Engineering Especially, and Engineering Education in General, L. T. Rader, 53

Research; Consulting engineers; Curricula; Education-practice interchange; Engineering education; Engineers; Model studies; Professional development; Professional engineering; Professional schools
Concept of a Professional School of Engineering, Charles H. Samson, 117

Research; Salaries; Computer applications; Computer graphics; Continuing education; Engineering education; Federal aid; Graduate study; Industries
Summary and Recommendations: Graduate Education in Engineering, Robert H. Page, 25

Research; Statistical data; Statistical distributions; Students; Teaching methods and content; Engineering education; Federal aid; Graduate study; Industries; Laboratory equipment
Effect of Federal Support for Programs and Students on Generation of New Faculty, Ross J. Martin, 109

Research; Statistical distributions; Students; Undergraduate study; Universities; Engineering education; Foreign engineering; Graduate study
Implications of Increasing Percentage of Foreign Nationals Enrolled in Graduate Programs for Engineering, W. Robert Marshall, 103

Research; Students; Women; Engineering; Engineering education; Engineering schools; Minority groups
An Examination and Discussion of Opportunities Available to Gifted Students with Engineering Interests — Problems and Interpretations, Lee F. Browne, 41

Research; Technical writers; Writing; Communication between engineers; Curricula; Engineering schools; Human factors; Information; Languages
Why are Engineering Students Not Being Taught to Write and Speak Well, and What are We Going to Do About It?, Ralph E. Jenkins, 69

Research; Undergraduate study; Education-practice interchange; Engineering education; Federal aid; Graduate study; Industries; Laboratory equipment
Summary and Recommendations: Research and Engineering Education, Richard C. Cunningham, 35

Salaries; Computer applications; Computer graphics; Continuing education; Engineering education; Federal aid; Graduate study; Industries; Research
Summary and Recommendations: Graduate Education in Engineering, Robert H. Page, 25

Salaries; Continuing education; Education-practice interchange; Engineering; Engineering education; Graduate study; Industries; Minority groups; Professional engineering
Summary and Recommendations: Engineering Practice and Engineering Education, Charles E. Schaffner, 29

Salaries; Sciences; Demand; Engineering education; Government; Industries; Mathematics; Minority groups; Motivation
Summary and Recommendations: The Pre-College Experience for Engineering Education, David R. Reyes-Guerra, 19

Salaries; United States; Universities; Engineering education; Government; Graduate study; Industries; Inflation (economic); Patents; Productivity
Crisis in American Engineering Education, Jerrier A. Haddad, 89

Sciences; Computer graphics; Computers; Cost effectiveness; Engineering education; Engineering schools; Industries; Man machine systems; Productivity
Effect of Computer Graphics in Both Industry and Engineering Education, Edward M. Rosen, 139

Sciences; Demand; Engineering education; Government; Industries; Mathematics; Minority groups; Motivation; Salaries
Summary and Recommendations: The Pre-College Experience for Engineering Education, David R. Reyes-Guerra, 19

Sciences; Social sciences; Undergraduate study; Curricula; Engineering education; Human factors; Humanities; Mathematics
Engineering Education and the Baccalaureate Experience (Engineering Education — The Liberal Education of the 80's), D. Allan Gianniny, 51

Sciences; Students; Teaching methods and content; Undergraduate study; Communication between engineers; Computer applications; Curricula; Engineering education; Humanities
Summary and Recommendations: Engineering Education and the Baccalaureate Experience, Russell R. O'Neill, 21

Social aspects; Sociological factors; Validation; Accreditation; Continuing education; Education-practice interchange; Industries; Licensing; Motivation; Professional engineering; Quality control
The Need for Continuing Education and Its Quality Control, Roy H. Matson, 137

Social sciences; Undergraduate study; Curricula; Engineering education; Human factors; Humanities; Mathematics; Sciences
Engineering Education and the Baccalaureate Experience (Engineering Education — The Liberal Education of the 80's), D. Allan Gianniny, 51

Sociological factors; Validation; Accreditation; Continuing education; Education-practice interchange; Industries; Licensing; Motivation; Professional engineering; Quality control; Social aspects
The Need for Continuing Education and Its Quality Control, Roy H. Matson, 137

Statistical data; Statistical distributions; Students; Teaching methods and content; Engineering education; Federal aid; Graduate study; Industries; Laboratory equipment; Research
Effect of Federal Support for Programs and Students on Generation of New Faculty, Ross J. Martin, 109

Statistical distributions; Students; Teaching methods and content; Engineering education; Federal aid; Graduate study; Industries; Laboratory equipment; Research; Statistical data
Effect of Federal Support for Programs and Students on Generation of New Faculty, Ross J. Martin, 109

Statistical distributions; Students; Undergraduate study; Universities; Engineering education; Foreign engineering; Graduate study; Research
Implications of Increasing Percentage of Foreign Nationals Enrolled in Graduate Programs for Engineering, W. Robert Marshall, 103

Subject Index

Students; Synthesis; Undergraduate study; Analysis; Comprehension; Education-practice interchange; Engineering schools; Evaluation; Graduate study; Logistics
Teaching of Engineering Courses with Student Faculty Ratios Far in Excess of Past Experience, John C. Lindenlaub, 65

Students; Teaching methods and content; Engineering education; Federal aid; Graduate study; Industries; Laboratory equipment; Research; Statistical data; Statistical distributions
Effect of Federal Support for Programs and Students on Generation of New Faculty, Ross J. Martin, 109

Students; Teaching methods and content; Undergraduate study; Communication between engineers; Computer applications; Curricula; Engineering education; Humanities; Sciences
Summary and Recommendations: Engineering Education and the Baccalaureate Experience, Russell R. O'Neill, 21

Students; Undergraduate study; Universities; Engineering education; Foreign engineering; Graduate study; Research; Statistical distributions
Implications of Increasing Percentage of Foreign Nationals Enrolled in Graduate Programs for Engineering, W. Robert Marshall, 103

Students; Women; Engineering; Engineering education; Engineering schools; Minority groups; Research
An Examination and Discussion of Opportunities Available to Gifted Students with Engineering Interests — Problems and Interpretations, Lee F. Browne, 41

Synthesis; Undergraduate study; Analysis; Comprehension; Education-practice interchange; Engineering schools; Evaluation; Graduate study; Logistics; Students
Teaching of Engineering Courses with Student Faculty Ratios Far in Excess of Past Experience, John C. Lindenlaub, 65

Teaching methods and content; Engineering education; Federal aid; Graduate study; Industries; Laboratory equipment; Research; Statistical data; Statistical distributions; Students
Effect of Federal Support for Programs and Students on Generation of New Faculty, Ross J. Martin, 109

Teaching methods and content; Undergraduate study; Communication between engineers; Computer applications; Curricula; Engineering education; Humanities; Sciences; Students
Summary and Recommendations: Engineering Education and the Baccalaureate Experience, Russell R. O'Neill, 21

Technical writers; Writing; Communication between engineers; Curricula; Engineering schools; Human factors; Information; Languages; Research
Why are Engineering Students Not Being Taught to Write and Speak Well, and What are We Going to Do About It?, Ralph E. Jenkins, 69

Undergraduate study; Analysis; Comprehension; Education-practice interchange; Engineering schools; Evaluation; Graduate study; Logistics; Students; Synthesis
Teaching of Engineering Courses with Student Faculty Ratios Far in Excess of Past Experience, John C. Lindenlaub, 65

Undergraduate study; Communication between engineers; Computer applications; Curricula; Engineering education; Humanities; Sciences; Students; Teaching methods and content
Summary and Recommendations: Engineering Education and the Baccalaureate Experience, Russell R. O'Neill, 21

Undergraduate study; Curricula; Engineering education; Human factors; Humanities; Mathematics; Sciences; Social sciences
Engineering Education and the Baccalaureate Experience (Engineering Education — The Liberal Education of the 80's), D. Allan Gianniny, 51

Undergraduate study; Education-practice interchange; Engineering education; Federal aid; Graduate study; Industries; Laboratory equipment; Research
Summary and Recommendations: Research and Engineering Education, Richard C. Cunningham, 35

Undergraduate study; Universities; Engineering education; Foreign engineering; Graduate study; Research; Statistical distributions; Students
Implications of Increasing Percentage of Foreign Nationals Enrolled in Graduate Programs for Engineering, W. Robert Marshall, 103

Union of Soviet Socialist Republics; United Kingdom; United States; Curricula; Engineering education; Germany; Graduate study; Japan
Key Aspects of Comparison Between Engineering Education in the USA and England, Japan, Germany, and Russia, J. S. Przemieniecki, 77

United Kingdom; United States; Curricula; Engineering education; Germany; Graduate study; Japan; Union of Soviet Socialist Republics
Key Aspects of Comparison Between Engineering Education in the USA and England, Japan, Germany, and Russia, J. S. Przemieniecki, 77

United States; Curricula; Engineering education; Germany; Graduate study; Japan; Union of Soviet Socialist Republics; United Kingdom
Key Aspects of Comparison Between Engineering Education in the USA and England, Japan, Germany, and Russia, J. S. Przemieniecki, 77

United States; Universities; Engineering education; Government; Graduate study; Industries; Inflation (economic); Patents; Productivity; Salaries
Crisis in American Engineering Education, Jerrier A. Haddad, 89

Universities; Curricula; Engineering; Forecasting; Management; Management engineering; Management systems; Management training; Professional engineering
Education for Engineering Management in the Eighties, Merritt A. Williamson, 157

Universities; Economic factors; Engineering education; Federal aid; Financing; Funding allocations; Graduate study; Laboratory equipment; Productivity; Recruiting
How to Increase the Number of Ph.D. Candidates and the Supply of New Faculty Members, Donald D. Glower, 97

Universities; Engineering education; Foreign engineering; Graduate study; Research; Statistical distributions; Students; Undergraduate study
  Implications of Increasing Percentage of Foreign Nationals Enrolled in Graduate Programs for Engineering, W. Robert Marshall, 103

Universities; Engineering education; Government; Graduate study; Industries; Inflation (economic); Patents; Productivity; Salaries; United States
  Crisis in American Engineering Education, Jerrier A. Haddad, 89

Validation; Accreditation; Continuing education; Education-practice interchange; Industries; Licensing; Motivation; Professional engineering; Quality control; Social aspects; Sociological factors
  The Need for Continuing Education and Its Quality Control, Roy H. Matson, 137

Women; Engineering; Engineering education; Engineering schools; Minority groups; Research; Students
  An Examination and Discussion of Opportunities Available to Gifted Students with Engineering Interests — Problems and Interpretations, Lee F. Browne, 41

Writing; Communication between engineers; Curricula; Engineering schools; Human factors; Information; Languages; Research; Technical writers
  Why are Engineering Students Not Being Taught to Write and Speak Well, and What are We Going to Do About It?, Ralph E. Jenkins, 69